D1327541

BALLISTICS IN THE SEVENTEENTH CENTURY

BALLISTICS IN THE SEVENTEENTH CENTURY

A STUDY IN THE RELATIONS OF SCIENCE AND WAR WITH REFERENCE PRINCIPALLY TO ENGLAND

BY

A. R. HALL, M.A., PH.D.

University Assistant Lecturer in the History of Science

CAMBRIDGE
AT THE UNIVERSITY PRESS
1952

PUBLISHED BY
THE SYNDICS OF THE CAMBRIDGE UNIVERSITY PRESS
London Office: Bentley House, N.W.1
American Branch: New York
Agents for Canada, India, and Pakistan: Macmillan

Printed in Great Britain by The Carlyle Press, Birmingham, 6

CONTENTS

LIST OF PLATES

Plates I, III and IV are reproduced by permission of the Syndics of the University Library, Cambridge. Plate II is a photograph of instruments in the Whipple Museum of the History of Science, Cambridge.

PREFACE

This essay was written with two purposes. The first was comparatively simple, to relate the foundation in the seventeenth century of the science of ballistics as a mathematical interpretation of some physical problems upon a basis of exact experimental investigation. This field is limited, both because a scientific analysis of these problems must rest upon a certain level of knowledge in pure mathematics, and because although references to the 'theory of projectiles' in contemporary works are frequent, the number of important contributions to its study is small. Secondly an attempt was made to relate this series of researches in applied mathematics to its background, social, economic and scientific, as a contribution to the history of science in society. Ballistics has been taken as a critical example of early science being directed by utilitarian considerations. My aim was to examine the justice of this suggestion more thoroughly, and to see how far such progress as was made could rightly be imputed to governmental interest, to the experience of a warlike age, or to the not uncommon feeling that the fruits of knowledge ought to be useful. In general I have been led to the opinion that the purposeful application of science to the art of war (and, it may be, to any technique or useful art) at any period before the nineteenth century is much less than at first appears from non-professional accounts; the conservative traditions of practical men yielded very slowly to the enthusiasm of inventive amateurs, whether scientists or not.

The work of the French military expert, Prosper Charbonnier, *Essais sur l'Histoire de la Balistique* (Paris, 1928) was of course a natural starting-point; that of Professor H. J. Tallqvist, *Översikt av Ballistikens Historia* (Helsingfors, 1931) was for an obvious reason less accessible to me. Both of these histories summarise briefly the major early treatises, but in a primarily mathematical treatment they ignore many questions of a wider historical interest. Perhaps this may justify a fresh research among the same, and other, materials.

It proved necessary to devote more space to military affairs than would otherwise have been desirable, for the reason that the history of war in the seventeenth century has yet to be written. The serious study of naval history has begun, for example in the publications of the Navy Records Society, but tactics and strategy still dominate military history. Some of the discussion which ranges outside science may therefore need to be modified in an ample consideration of the technical and economic history of war.

In a slightly different form this essay was accepted by the University of Cambridge as a doctoral dissertation. I owe much to the advice of the Provost of Oriel College, Oxford, and to Mr F. P. White of St John's College, who supervised its progress: but are not responsible for its faults; and to Professor Butterfield and others who were kind enough to read and criticise. I wish also to express my obligation to the Royal Society for permission to explore its archives, and to Lord Courthope of Whiligh for placing family papers at my disposal. I am deeply indebted to my College and University for the support which made this work possible, and to the Syndics of the University Press for undertaking its publication.

A.R.H.

CAMBRIDGE
February 1951

CHAPTER I

THE BACKGROUND OF SCIENCE I GOVERNMENT AND INDUSTRY

Science and Technology

To say that the economic life of society in general, and processes of manufacture in particular, were unaffected by science until the beginning of the last century is scarcely an exaggeration. The seventeenth-century revolution in thought and method had moulded a science which was potentially capable of effecting profound changes in the means of production, and in fact many writers on science at the time found an important justification for the study of science in the fuller exploitation of natural resources, with the consequent enrichment of human life and alleviation of daily toil which it promised. But this promise was only fulfilled through the industrial and agrarian revolutions of the nineteenth century and the changes in the organisation of economic activity which they brought about. In particular the sudden rise of engineering needs above the level of the carpenter and the black-smith, the sudden realisation that engineering skill in all its branches was fundamental to improvements in manufacture, transportation, agriculture and the means of making war, created a situation in which scientific knowledge and method not only could be, but must be, applied, while large-scale manufacture provided the means and incentive for the application of science.

It is easy to see why the hopes of men like Hartlib, Petty and Boyle, who had written at length of the social usefulness of science, had failed of immediate realisation. The attainments of seventeenth-century science were very great, but in matters of detailed explanation it was still weak. Even the fundamental principles of chemistry, of the science of living nature, of the science of the earth, were still wanting. Outside mathematical science the natural philosopher had very little to offer to the craftsman; even where he suspected that traditional methods were wasteful and inefficient, he could not suggest a remedy. Although a few professions—surveying, navigation and instrument-making, for instance—were now based upon a firm mathematical foundation and many others were undergoing modification in the direction of scientific method, the staple occupations of mankind —agriculture, mining, the cloth industry, ship-building—were the province of tradition and craft lore, unshaken as yet by the questionings of science.

If, in studying early science, we must free ourselves from the notions of 'pure' and 'applied' science which are little more than a century old, if we find that the distinction between science and technology becomes vague, we must also recognise that this was but an aspect of a more general broadness of thought. The natural philosophy of the seventeenth century, while it knew such principal divisions as mathematics, physics, chemistry and astronomy, allowed a free intercourse between them. It was still possible for the assiduous student to embrace the whole of science in his mind, and fit himself to appreciate every important event in the intellectual history of Europe. Huygens, for instance, was not only a master of physical science, but made discoveries in each of its branches. In Newton—physicist, mathematician, chemist, theologian, economist and public servant; in Leibniz—mathematician, philosopher, historian and politician; in John Ray—biologist, botanist, theologian, philologist—we have examples of the fact that there was no breach between science, scholarship, and the world of affairs.

The diversity of interest and the power of such men as these was not unusual, for if every man of learning was not a natural philosopher, every scientist was a man of learning, and many of them lived in the 'great world'. The science they created was marked by the absence of professional narrowness. Apart from physicians, the only professional scientists were those in universities and similar institutions engaged in the teaching of natural philosophy and mathematics. The typical seventeenth-century scientist was a gentleman who, if he was unable to live on his income, entered upon the ministry of religion, the practice of medicine, or the service of the state. By the standards of a later age he was an amateur and dilettante, unconscious of a deep distinction between science and his many other occupations. He turned with equal interest from a mathematical theorem to a prodigy of nature, from travellers' tales to a trade secret. Natural philosophy was not so much a new compartment of learning—though this, as it hardened into 'science', it became—as a new way of approaching and acquiring knowledge. The collection and consideration of a type of fact which had been neglected in all previous human experience was less important than the new attitude of mind in which the task was begun.

So far was the scientific mind of the seventeenth century from observing an artificial distinction between pure and applied science —mixed science, as Boyle called it—or between science and technology, that it was a common-place of the time to explain how science could profit from the experience of tradesmen, and trades from the teachings of science. The chemists, to profit by the experience of metal-workers, studied Agricola's *De re metallica* and Biringuccio's *De la Pirotechnia*, the two great German and Italian masterpieces on the art. The Royal Society collected histories of trades, in which the whole process used in many crafts was recorded. Galileo offered the advice that mechanics might profitably be studied in the Arsenal at Venice. And in the words of Robert Boyle, who was at once a great physicist, a great chemist, a great interpreter of science, and a weighty writer on divine subjects:

> The Phenomena afforded us by these [mechanical] arts ought to be lookt upon as really belonging to the history of nature in its full and due extent. And therefore as they fall under the cognizance of the naturalist and challenge his speculation, so it may well be supposed that being thoroughly understood they cannot but much contribute to the advancement of his knowledge, and consequently of his power, which we have often observed to be grounded upon his knowledge and proportionate to it.[1]

It would be mistaken to suppose that science had ever been confined strictly within the cloister and the closet, but it is

[1] *Some Considerations touching the Usefulness of Natural Philosophy* (Oxford, 1664): 'Of the Usefulness of Mechanical Discipline, 2'. cf. John Wilkins, *Mathematical Magick*. (London, 1648), Preface.

certainly true that there was now a strong movement to draw it closer to the field, the forge, and the workshop.

Following the contemporary interpretation, we may say that the divine studied science in order to understand the works of the Creator, the philosopher in order to satisfy his intellectual curiosity, the craftsman, by instinct, in order to turn his knowledge to useful ends. None of these motives was necessarily more noble than another, and there was indeed a strong link between the first and the last, for it was universally believed that the empire of man over nature had been the divine intention, and that nothing in the creation was without some purpose or lesson for man. If many of these lessons were not yet apparent, it was because science had been neglected and the investigation of nature scarcely begun. These were not new ideas, but they received greater force, and utility became a respectable reason for the study of science. To reduce nature to the status of a machine which could be managed for man's purposes did not seem incompatible with the view of nature as the supreme example of the wisdom and beneficence of God. Yet, for the reasons already given, the progress of science was little influenced by the utilitarian outlook of some scientists, although the practical possibilities of science recommended it to commerce and government. Advancing knowledge severed rather than strengthened the links between botany and medicine, astronomy and navigation, mathematics and mensuration. The synthesis of natural philosophy and technology which had been a principal object of the Royal Society, lauded by publicists and welcomed by statesmen, was never achieved, perhaps because the real intellectual vigour of the scientific renaissance sprang from far deeper roots than they realised, and was to work a more fundamental change in the human scene than they could have anticipated.

Robert Boyle, the eloquent exponent of the 'Usefulness of Natural Philosophy', has also the distinction of being the first Englishman to use the word 'ballistics' in print in the work to which he gave that title, where he classes ballistics with pneumatics and hydraulics as a mixed science.[1] He was justified in his definition, for ballistics was the fruit of the scientific revolution which had been effected by the preceding generation, and the

[1] See below p. 78.

most finished of Galileo's two new sciences, applying the philosopher's theory of projectiles to the practical needs of the gunner and the bombardier. As a science, ballistics was studied by many of the heirs of Galileo, the mathematical physicists of the seventeenth century, until it entered upon its modern history in Newton's *Principia*; while as technology it had begun to penetrate into practical manuals of instruction and give a scientific veneer to the art of gunnery. For the scientists there was no need to explain their interest in the problems of ballistics, since they arose naturally in the course of their investigation of motion, the most fruitful and most complete of all early researches. Moreover, the solution of these problems required the highest extension of mathematical methods. If at one extreme ballistics touched upon the crude equipment and simple artifices of the gunner, at the other it played an important part in the working out of the laws of mechanics.

To put the point more generally, in ballistics physical science, technology and gunnery, which is a branch of the art of war, combine. The relations of science and the art of war are similar to those of science and technology. Indeed, from the point of view of statesmen they are identical, for the power and skill of the state in making war depend (among other things) upon its technological development, and this in turn depends upon the application of science. This truth had been dimly apprehended in the late sixteenth century, and in the age of Newton was too clear to require demonstration. It was apparent to Colbert when he founded the Académie Royale des Sciences and it was urged by Leibniz as a reason for founding a similar scientific assembly in Germany.

The requirements of military supply, in a century of almost incessant warfare (in the course of which, moreover, many new, expensive and complicated means of destruction were invented) pressed heavily upon industrial resources. To the mercantilist statesman no object was more desirable than national self-sufficiency in armaments. Scientific knowledge could assist in attaining it by the discovery of new processes and the substitution of native commodities for those formerly imported.[1] Elizabeth of England

[1] J. U. Nef, 'War and Economic Progress, 1540-1640'. *Ec*[*onomic H*[*istory*] *R*[*eview*] vol. XII (1942), p. 19.

took a benevolent interest in the progress of new ventures into mining and metallurgy, looking to them for a secure supply of brass for ordnance. For the same reason in addition to cheapness, iron ordnance were used rather than brass wherever possible, though the latter metal was usually said to be superior. The native saltpetre industry (despite the hostility of the country to the digging up of the floors of its stables and pigeon lofts) and researches into the chemistry of explosives were encouraged by English sovereigns until the East India Company's trade provided a cheap and safe source of supply of this essential chemical.[1] Prince Rupert, cousin of Charles II and one of his ablest naval commanders, encouraged the Royal Society in their experiments with explosives and projectiles.[2] Colbert set the members of the Parisian Academy to work on direct military problems, Huygens examining the traction of gun-carriages, Blondel, Roemer and De la Hire exterior ballistics.[3] To multiply instances would only suggest that the science of the seventeenth century was more practical and more military than in fact it was; these suffice to show that the history of ballistics can only be studied by considering four separate topics: technology; the relations of government and industry; gunnery; and scientific ballistics. The first two of these topics will be treated in the remaining sections of this chapter; the third in the next, and the fourth in the remaining chapters.

2. *The Manufacture of Artillery*

It is a fairly obvious cause of error in assessing the work of early scientists to suppose that, because they were not unaware of the practical value of some of their discoveries, whenever they hit upon some truth capable of a technical application it was necessarily of technological importance. Huygens found the way to regulate a clock by means of a pendulum, and thereby revolu-

[1] *C[alendar of] S[tate] P[apers], D[omestic], passim*; H. A. Young, *East India Co.'s Arsenals* (Oxford, 1937), pp. 62 *et seq.*

[2] Thomas Birch, *History of the Royal Society* (London, 1756–7), *passim.*

[3] *Œuvres Complètes*, tome. XIX, p. 48; *Histoire de l'Académie Royale des Sciences* (1733), tome I, p. 71. Papin, who derived from Huygens much of his scientific knowledge, read a paper on the same subject before the Royal Society in 1711 (*Journal Book* vol. X, p. 315). François Blondel, *L'Art de Jetter les Bombes* (Paris, 1683). Roemer's notes on the proof of cannon, etc., are in *Adversaria*, pp. 191 *et seq.*

tionised the clock-making industry and the habits of our ancestors; but Roemer's discovery of the optimum shape for the teeth of gear-wheels did not have a corresponding effect upon the design of machinery.[1] Before it is possible to declare that a certain scientific investigation was of technological importance, though it may clearly appear to later eyes to be so, before we can declare that the investigator was guided by motives of practical usefulness, we must know whether the state of technical knowledge was such that the scientific principle in question could be applied. It was profitable to the clock-maker to make Huygens' new clock; it was not profitable to the millwright to make his gears in accordance with Roemer's theorem. Before we can say how far the scientific ballistics of the seventeenth century affected gunnery, it is needful to know how guns were made and of what they were capable. Did the study of ballistics arise out of an interest in particular problems of mathematical physics, or out of governmental pressure? Were the mathematical theories relevant for practical men? Was there—to raise the general question—any relation between science and war in the seventeenth century? Such questions cannot be answered from the history of science alone, for the putting into practice of the military potentialities of science was always the work of soldiers and engineers.

The development of weapons of war has been the result of a balance between the requirements of military enterprise, the ingenuity of inventors, and the capability of artisans. If we may judge from the military writings of the seventeenth century, the soldier was tolerably well satisfied with the weapons he had, and did not eagerly welcome innovations. In this he was supported by government, which had no desire to increase the already heavy burden of military expenditure. In so far as science was supported at all for military purposes, the chemist who promised more powerful explosives and incendiary materials was favoured, rather than the ballistician who offered greater accuracy. Military conservatism and the fact that war concerned only a very small portion of the population spared Europe the development of its

[1] This point, and the whole question of why early science failed to influence technology, has been discussed by G. N. Clark, *Science and Social Welfare in the Age of Newton* (Oxford, 1937), ch. I.

ordnance from the sixteenth to the nineteenth century. From the invention of cast bronze cannon in the late fifteenth century to the disappearance of the smooth-bore, muzzle-loading gun about a century ago, there was no radical change in the design of artillery. Manufacturing methods altered. Iron was introduced as a gun-metal in the mid-sixteenth century; the scale of cannon increased in the seventeenth; better boring methods were used in the eighteenth; but the guns of Queen Victoria's wooden ships were capable of little more accurate practice than those of Drake's fleet which defeated the Armada.

In this respect there is a great difference between the history of hand firearms and the history of artillery in our period. The military experts invariably, the scientists whenever they applied their hypotheses in concrete cases, wrote of the ballistics of cannon and large projectiles. The ballistics of small-arms was entirely neglected, for the state of manufacturing technique forbade their use at any range beyond the point-blank 150-200 yards.[1] Yet, through various inventions of a technical character, the hand firearm passed through a period of rapid development from about 1550 to 1650, after which it scarcely changed for almost two hundred years. None of these inventions, however, had the effect of modifying tactics or of creating a new scientific interest, since none of them increased the effective range or accuracy or the weapon. The sporting rifle, a type of gun which did much to popularise the new sport of shooting—as we may see from the *Life* of Benvenuto Cellini—was the only accurate gun. This 'fowling piece', elaborately chased and ornamented to suit the quality of its possessor, was far too precious to be placed in the ignorant hands of the common soldier, on account of the many hours of highly skilful smith's work with forge and file which had gone to its manufacture.

The military inventor was less fertile in suggesting improvements in the manufacture or use of artillery, and military departments were not so easily persuaded to adopt them. The commander in the field was less free to experiment, for the artillery was directly under the control of the Master of the Ordnance, who was

[1] There were of course discussions by military writers of the best weight, size and calibre of arms for different types of service, and of the appropriate charges of powder. See, e.g., *C.S.P.D. 1638-9*, p. 189; *1639-40*, p. 398.

accountable for every piece in fort, store or aboard ship. The trial-and-error methods which brought the gunsmith to better his skill by insensible degrees were not available to the cannon-founder; too much was at stake, in the loss of capital and the prejudice of operations. Since every hundred pounds of metal in a piece of ordnance cost about as much as a musket, even a train of field guns represented a sizeable capital investment which was expected to last for many years.[1] Though the Crown made large payments to the London Gunmakers (incorporated in 1637),[2] the usual suppliers of small-arms, the capital investment and yearly production of the gun-foundries of Kent and Sussex had a far greater value. Thus while small-arms were the product of skill at the bench on a relatively slight weight of material, which could be applied to new designs as required, gun-founding was a highly organised and costly manufacture, whose techniques were already stabilised and would not easily admit of modification even to cast a better type of gun. The smith is naturally more adaptable than the founder. However, in practice the latter was rarely required to alter his accustomed methods, since the design of artillery remained essentially unchanged, less on account of the inability of science and industry to produce better weapons than of the absence of pressure to this end. Military opinion was happy in a surviving tradition of chivalry that close combat was more honourable than a long range bombardment between invisible foes.

Artillery began to change the character of warfare in the second quarter of the fourteenth century.[3] For a century even large guns, like the famous Mons Meg at Edinburgh, were built by the smith from bars and hoops welded into a crude tube, several such sections being screwed together for greater length. Then the blast-furnace made possible the pouring of large quantities of metal. This new technique, with the cast bronze guns it produced, was introduced into England by the early Tudors, who brought over foreign workers to establish the new industry, just as Edward III

[1] Amid wide short-term fluctuations in prices of military stores, it may be assumed that in the middle of the century guns, by gross weight, were valued at about £20 a ton (iron) and £30 a ton (brass); small-arms could be bought for 15-20 shillings apiece.

[2] The Charter is printed in the *Journal of the Society for Army Historical Research*, vol. VI (1927), p. 79. The Gunmakers had petitioned for a charter as early as 1581, and it was finally granted to them in the interests of national defence.

[3] Sir Charles Oman, *The Art of War in the Middle Ages* (London, 1924), pp. 205 *et seq.*

had hired foreign smiths to make the small iron guns used at Crécy; and, as before, Englishmen soon acquired their skill. In 1543, pressed by financial difficulties which prevented the purchase of brass abroad, Henry VIII sent a French founder from the royal foundry of bronze ordnance at Houndsditch to a Sussex iron foundry, where 'gun-stones' (round shot) were already being cast, and there Peter Baude poured the first iron guns in a single casting, probably applying to molten iron the very techniques of brass-casting which Vannoccio Biringuccio described in *De la Pirotechnia*, published three years before. From this technological revolution the English iron ordnance industry was born.[1]

Though the tutelage of England in metallurgy continued for at least a century and a half, the Sussex iron industry was of immense economic and political significance. It placed a powerful source of military strength in the hands of the commercial, Protestant powers; it gave Englishmen their first sense of industrial importance. But there is no evidence that the manufacture of iron cannon in any way differed from that of brass ordnance, or that Englishmen introduced any innovations.[2] There is indeed no detailed contemporary account of the English iron ordnance manufacture, so that the published descriptions of Biringuccio (1540) and Saint-Rémy (1697), referring to Italian and French practice, with a Spanish manuscript by the navigator Diego de Prado y Tovar (1603), must be relied upon as our only complete literary records.[3]

Comparison of the accounts by these writers, substantiated by fragmentary references to English methods, suggests that the technique of ordnance-casting never advanced far beyond the primitive skill of the bell-founder; that the process known to Biringuccio was exactly that described by Prado y Tovar and Saint-Rémy, so that there were no important variations of time and place in our period; and that it was entirely unaffected by the

[1] Ernest Straker, *Wealden Iron* (London, 1931); Rhys Jenkins, 'The Rise and Fall of the Sussex Iron Industry,' *Transactions of the Newcomen Society*, vol. I (1920), p. 16; H. Schubert, 'The First Cast-Iron Cannon made in England', *Journal of the Iron and Steel Institute*, vol. CXLVI (1942), p. 131.

[2] The seventeenth-century patents for smelting with coal may, of course, be disregarded. This point is also touched upon in the conclusion of this essay.

[3] Surirey de Saint-Rémy, *Mémoires d'Artillerie* (Paris, 1697); Diego de Prado y Tovar, *Encyclopaedia de Fundicion de Artilleria y su Platica Manual*, 1603 (MS. in Cambridge University Library).

scientific revolution. Metallurgy was extensively studied by the early scientists, and the chemistry of metals, from the point of view of the alchemist or the physician, formed an important section of most of the seventeenth-century chemical treatises. But practical metallurgy was entirely derived from the metal-workers themselves.[1] Neither then, nor for a very long time, was a scientific explanation of such elementary operations as the reduction of ores to metal, the formation of alloys, or the conversion of cast iron into wrought iron and steel possible. Experiments were made by gun-founders, in the absence of scientific knowledge relying on craft, skill and empiricism, to discover closer and finer alloys of tin, zinc, and copper less subject to porosity in casting, without success.[2] Prince Rupert took out a patent for a way of annealing and turning cast iron cannon about 1678, intending thereby to produce a tougher and better finished gun. It is possible that the Prince had made a really significant discovery—that cast iron may be softened through the reduction of its carbon content by heating with an oxidising agent such as iron oxide or bone ash—but his new ordnance were too expensive to compete with ordinary iron and brass guns.[3]

The most usual types of cannon—although a small number of breech-loading port-pieces and swivel guns useful at close quarters were still retained in the Navy—were cast hollow in a single massive block, in just the same way as the ordinary kitchen pot. The art of the process lay in shaping the mould correctly and in reducing the molten metal to the necessary fluidity and fineness, so that the mould should be completely and densely filled when the furnace was tapped. The quality of the piece depended entirely upon the accuracy of these operations, for the subsequent machining could not correct a defective casting.[4] The first task was to

[1] E.g. John Webster, *Metallographia or an History of Metals* (London, 1671).

[2] P. Boissonade, *Colbert, 1661–1683* (Paris, 1932), p. 102; Saint-Rémy, *op. cit.* vol. II, pp. 45 *et seq.*

[3] Eliot Warburton, *Memoirs of Prince Rupert and the Cavaliers* (London, 1849), vol. III, p. 494. Samuel Pepys, *Naval Minutes* (N[avy] R[ecord] S[ociety], 1925), pp. 51, 225. The guns were annealed in glass houses and bored by water-power; Réaumur described the process of softening cast iron in *L'Art d'Adoucir le Fer Fondu* (Paris, 1722). I have not discovered whether the Prince knew of the importance of a decarbonising agent. Cf. also W. H. Hatfield, *Cast Iron* (London, 1918), pp. xiii, xvii.

[4] The importance of the correct shaping of the mould is particularly stressed by Gaya, *Traité des Armes, des Machines de Guerre* (Paris, 1678), liv. II, cap. I.

build up with clay on a wooden frame a replica of the outside of the cannon to be cast, to which the decorations, trunnions, cascabel, etc. were attached by pins. On this the mould proper was built up, the replica being covered with a mixture of ashes and fat to prevent the clay of the mould sticking to it. The clay was added thinly at first to take up the outlines of the replica and give a smooth finish, then more thickly, and finally the mould was bound with iron rods and wire and left to dry.

Next the wooden frame was drawn out, the replica broken up and removed with long bars, and the attachments taken out of the mould, which was thoroughly baked. Within this mould, which now retained all the exterior lines of the gun, a core of clay, modelled round an iron bar to the shape of the bore, was set in position by means of iron supports in the mould, leaving a space equal to the thickness of the metal in the casting. The complete mould was lowered breech-first into a pit close to the furnace mouth and the metal run into it. When cool the casting could only be got out by breaking up the mould, so that the whole process had to be repeated for every piece, and no two could be cast exactly alike.[1] The cannon was finished by cutting off the waste metal at the mouth, drilling the touch-hole, smoothing the outside with file and chisel and reaming out the bore, this last a most important operation since it gave polish to the interior of the casting, but the reamer was incapable of correcting any faults in the alignment of the bore (due to misplacing the core of the mould) or of converting the bore into a true cylinder. Early machines for this purpose were very crude, but Prado y Tovar draws a picture of an efficient vertical boring machine driven by power. Probably, however, the early iron guns were not bored at all, but used in the rough state.

The gun was now ready to be proved, to show whether it was

[1] Sir George Carew, Lieutenant of the Ordnance, wrote to Cecil (31 July 1594) that 28 cwts. was the normal weight of a ship's demi-culverin, but that the founders never cast so exactly but their pieces varied 2-3 cwt. (*C.S.P.D. 1591-4*, p. 532.) In the 1660's guns were taken into the Tower store from the foundries of Alexander Courthope varying in weight as follows: Demi-cannon 42-7 cwts., Culverin 34-8 cwts., Demi-culverin 30-2 and 24-6 cwts., Sakers 11-12 cwts., so that Carew did not exaggerate. This was much improved by the mid-eighteenth century when John Fuller could write of two small guns that each was within three pounds weight of the model and within three or four of the other's weight (*S[ussex] A[rchaeological] C[ollections]*, vol. LXVII, p. 49).

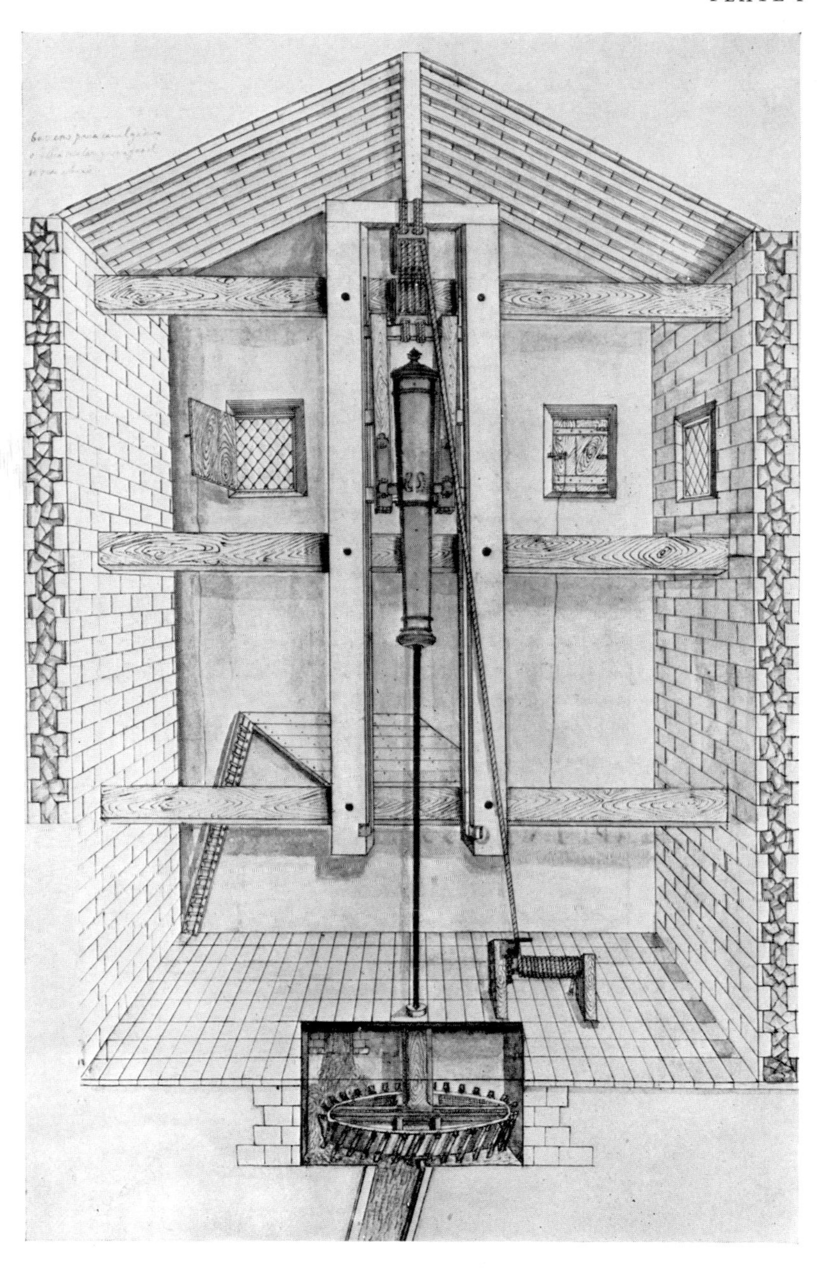

A MACHINE FOR REAMING OUT THE BORE OF CANNON, 1603

sound and fit for service. In England no cannon or firearm of any sort was admitted to the Tower armouries until it had passed the prescribed test. The method of proof by which a gunner should satisfy himself that his weapon would not blow up on firing the first round at the enemy was described by most of the writers on artillery, varying very little in the course of the century.[1] The first examination was physical: the gun was well viewed to ascertain that the metal was a good alloy, not too soft or too brittle, and that the bore was of the proper diameter with due allowance for 'windage' beyond the calibre of the ball, and free from holes and cracks indicating weak places in which, moreover, a smouldering fragment might lodge to destroy the gunner when next he inserted a ladleful of powder. The gunner would further make sure that the trunnions were symmetrically placed with respect to the axis of the piece, and that the bore was parallel to it, for otherwise it could never shoot straight. After passing this inspection, the gun was fired with ball and three increasing charges of powder, the last equal in weight to the shot. If the gun withstood this test without bursting it was accepted as safe.

The practice in the French proof in the later seventeenth century was to give the large charge first, and two smaller ones after. The gun was then searched, filled with water, and a plunger worked in the bore to force the water through any cracks.[2] This was the Dutch system, only imitated in France during the reign of Louis XIV. The test imposed by the Board of Ordnance in England seems to have been more severe, for references to the King's double proof indicate that the proving charge was as much as four-thirds the weight of the ball.[3] No cannon, on any system of proof, was required to undergo more than a rough test of

[1] See, e.g., Gabriel Busca, *Della Espugnatione et Difesa della Fortezza . . . Aggiontari nel fine l'instruttione de Bombardieri* (Turino, 1598).

[2] Saint-Rémy, *op. cit.* vol. II, p. 71. Georges Fournier, *Hydrographie contenant la Théorie et la Pratique de toutes les Parties de la Navigation* (Paris, 1643, 1667), liv. III, cap. XIV, XV.

[3] *C.S.P.D. 1625–6*, p. 30. W.O. 47/19A, p. 46. In a paper written by Sackville Crowe (Wallace Notestein *et al.*, *Commons Debates of 1621*, vol. VII [1935], p. 168) it is explicitly stated that 'the trial [is] to be with a double charge to what they usuallie shoote withall'. Brass ordnance was less keenly proved than iron. An entry in the ordnance accounts for 1637 (*Ordnance Minutes 1636–1639*) under date 13 Feb. gives charges for proof and service 'by advice and opinion of Mr Thomas Pitt his Majesty's founder of Brasse Ordnance' as follows: Cannon of 7, 24 and 18 lb., Demi-cannon 18 and 15 lb., Culverin 14 and 10 lb., etc.

strength and workmanship, and its accuracy in gunnery was unassessed.

In the Wealden iron industry, which in addition to the normal domestic trade in iron ware supplied the Board of Ordnance with all the cannon it required, furnished merchant shipping, and carried on a flourishing export trade throughout a large part of the seventeenth century, the unit of production remained small.[1] Ordnance were cast in works which had only a single furnace, capable of smelting about 200 tons of metal in a year, if we reckon about 6–8 tons as the yield per found-day or working week, and 25–35 found-days in the year. Great arsenals or factories like that of the Duke of Lorraine at Nancy, covering several acres of ground and as populous as a large village, were unknown in England, and uncommon anywhere.[2] Usually the permanent installation—furnaces, hammers, forges and chaferies—was provided by the landowner and leased by the ironmaster, who as entrepreneur provided ore and fuel for the founder whom he employed, paying him by the working week. The ore, wood, and loam for moulds were bought by the load at the furnace, for it was left to neighbouring landowners to make what they could of their own resources. The Sidney accounts of the sixteenth century agree with the Fuller accounts of the eighteenth in representing the ironmaster as an organiser of several trades towards a common end, the bricklayers and carpenters employed about the furnace and machinery, the miners and coal-men who brought iron-ore and charcoal, the moulder whose skill was the key to all, were paid by the task as required. Similarly the transportation of finished guns to the ports—whence they went up river to Tower wharf—was let out by contract. The founder and his 'fillers', busy as long as the furnace was in blast, were the only permanent employees of the ironfounder, and there was a wide division between him, as business manager and a gentleman, and

[1] The operation of the furnace is described by John Ray in *A Collection of English Words not generally used . . . with an account of the preparing and refining such metals and minerals as are gotten in England* (London, 1691). In *Emmanuelis Swedenborgii Regnum Subterraneum sive minerale de Ferro* (Dresdae et Lipsiae, 1734), p. 158, there is a short chapter on 'Fornaces ferri fusoriae pro tormentis bellicis Angliae', which beyond a reference to the double furnace, and the casting of guns in winter, pig-iron in summer, is of little interest.

[2] Nef, *loc. cit.* pp. 22 *et seq.*

the workmen in whose hands the success of the undertaking really rested.

The activity of the furnace was very variable since it depended greatly on the vagaries of the weather as well as on the state of the market. On some sites the summer rainfall was insufficient to fill the streams providing power for the blast, and for these the summer was a slack season, while in winter it was impossible to move the finished guns. The technical and commercial problems of gun-founding were such that the founders might well protest at hasty modifications of pattern, and be impatient of the ignorance of the Ordnance Board of the economics of the industry.[1]

There are obviously two aspects of the mechanisation of weapons, the building of machines and the using of machines. The important part played by the development of weapons in the history of the machine has been emphasised by Lewis Mumford, who writes 'Within the domain of warfare there has been no psychological hindrance to murderous invention, except that due to lethargy and routine: no limits to invention suggest themselves.'[2] But in making new inventions that really worked, the practical craftsmen, the empirical engineer and the professional soldier took the lead; the scientific inventors sketched brilliant conceptions—Leonardo's war-chariots, the submarine of Cornelius Drebbel, the devices of Napier of Merchiston—which were absurdly inappropriate for the prevailing techniques of manufacture and warfare. Not being engineers or soldiers they neglected the difficulties which would have been apparent to the carpenter or the smith, and indeed, it was the new *use*, rather than the manufacture, which exercised their imaginations. In the conditions of early science the machine was only considered in so far as it was perfect for its intended use, the compass to draw a perfect circle, the air pump to create a vacuum, the pulley to multiply forces. In the same way the gun was treated by early ballisticians from

[1] Sidney accounts *H.M.C. De l'Isle and Dudley*, vol. I, pp. 311-12, etc. Fuller accounts *S.A.C.*, vol. LXVII, pp. 27-8, 31, 39, 42. One of the ordnance officers writes to George Browne (31 Jan. 1665/6) telling him to warn his workmen not to cast brass Cannon of Seven 'Gouty & heavy at ye Muzzle, but the Mettle runne to ye ffortiffieing of ye peece at ye Chamber' (*Courthope Papers*).

[2] *Technics and Civilization* (London, 1946), p. 85. This is scarcely fair to scientists who suppressed some of their own murderous inventions 'under the restraint of humane ideals', as he acknowledges later on the same page.

the point of view of *use*, not manufacture, as a perfect engine for projection. Neither Galileo, Torricelli, Huygens nor Newton investigated the technology of the ordnance industry or the irregularities of gunnery, and if they had, their researches could scarcely have been affected. The student of exterior ballistics has to treat the gun as a perfect machine, since his work would otherwise be deprived of all basis, but the usefulness of his work will rest entirely upon the extent to which technology is capable of producing a gun corresponding to this ideal.

In all respects the cannon of the seventeenth century were imperfect. They were individual, whereas the first necessity of scientific gunnery is that every gun shall be indistinguishable from its fellows of the same type. The accuracy of every piece depended entirely upon the individual judgement of the workman and the military standard was such that every gun, however inaccurate, was fit for service as long as it would stand a charge. The standard of engineering technology was not merely insufficient to make scientific gunnery possible, it deprived ballistics of all experimental foundation and almost of the status of an applied science, since there was no technique to which it could, in fact, be applied.

3. *The Military Department of the State*

In England, as in France and Spain, the responsibility for the administration, supply, effectiveness and therefore ultimately the design of the royal artillery rested upon the Master of the Ordnance, whose office appears to be almost coeval with the invention of artillery. In practice the Master was sometimes no more than a titled figurehead, an intermediary between the Privy Council, the makers of policy, and the Principal Officers of the Ordnance, who were familiar with the necessary technical details.[1] When guns were to be bought or ordered out of the stores for service, the types required were sometimes decided by the Council on the advice of its military members, sometimes the Officers were directed to draw up an estimate of what would be required. In

[1] It also became unusual for the Master to serve as General of the Artillery in the field, though both Schomberg and Marlborough held the office while serving as commanding generals.

any case they were expected to be able to state from their ledgers and periodical surveys what stores were available within the Tower and on service-stations towards the fulfilment of any task, though the movement of every gun, every barrel of powder and every musket was strictly controlled by the Privy Council itself. All these things depended upon the weight, cost and expenditure of ammunition of each type of cannon as well as upon the sort of service to be expected from it in the field. It would have been ruinous to mount 6-pounders in a ship of the line, or to send demi-cannon with the field artillery.

The potent influence of the Board of Ordnance on tactics may be illustrated by an incident in the preparations for getting a fleet to sea in 1664, during the Second Dutch War.[1] The Admiralty proposed to increase the fire-power of the ships by mounting larger numbers of whole cannon, demi-cannon, and demi-culverins; but the Board was unable to supply them from its stores. Even on its own much reduced estimates it was necessary to cast over 500 new cannon of the more powerful types, though there was a large surplus of smaller guns. Thus, when the delay and expense of manufacturing ordnance are remembered, it was inevitable that the Ordnance Officers should have the final decision on the numbers and types of guns ordered for active service, and if they were to carry out their duties efficiently and avoid failures like this of 1664, it was essential that they should know not only the strategical policy of the Crown but the proper performance and supply of each gun whether on sea or land service. To enter into a discussion of details like these would occupy too much space and contribute little to the understanding of the problems of ballisticians. It will be sufficient to say that in the experience of commanders (reflected in the writings of the military experts) there had been built up a knowledge sufficient to decide which gun was most suitable for a particular purpose, what supplies and skilled management it would require.

Broadly speaking, the Board left the design and manufacture of cannon to the experience and ingenuity of the ironfounders' workmen, expecting no more than the equality of new guns with those already in service. The great majority of contracts and

[1] Public Record Office, War Office Records, W.O. 47/5, pp. 237 *et seq.*

orders in official records simply enumerate the type to be cast, whether whole cannon, demi-cannon, culverin, etc., and the length of the piece (for the larger guns were made in three lengths of $8\frac{1}{2}$, 9, or 10 feet). Some contracts specify the thickness of the metal at the touch-hole, the trunnions and the neck.[1] There are of course exceptions to this general rule. The ornamentation, coats of arms and so on were ordered by the Board as appropriate for the use to which the cannon were intended. It was occasionally necessary to specify whether they were to be 'drake' or 'whole-[home-] bored'. There are references to designs being supplied to the gun-founder by the Master Gunner of England.[2] It was natural, for instance, that the founder should be instructed, when casting eight three-pounder brass cannon for Charles II's yacht the *Henrietta* to make them 'all home bore according to the mouldings and dimensions already issued' to him.[3] An intervention of a different type survives in a Board minute of 7 Nov. 1668 that the iron Cannon of Seven cast by Thomas Westerne be taken off his hands and put in store, having survived a double proof in spite of a flaw, this piece having been 'cast by their order for an Experiment only'.[4]

As this minute suggests, the Crown instructed the Officers of the Board of Ordnance to test the value of inventions presented to it from time to time. Though neither the Board collectively nor any of its officers in their official capacities had the means or authority to carry on research of any sort, the Master Gunner and the Fire-master, who were its experts on gunnery and bom-

[1] It was well known that the longer gun had the greater range, and length was therefore an advantage to be balanced against unwieldiness (Albemarle, *Observations upon Military and Political Affairs* [London, 1671], p. 33). Several specifications of this type are preserved in the *Courthope Papers*.

[2] *S.A.C.* vol. LXVII p. 49 (September 1646). Blake writes (13 Feb. 1652/3): 'whether [the cannon contracted for the fleet] be made drakes or home-bored it is our advice that, provided they be made of the same weight and yet allow the same metal as you do for whole bored guns, drake-bored will be of most use, otherwise to make them whole-bored, (W.O. 47/2, unpaged). The *O.E.D.* defines 'drake' as a small type of cannon. This is clearly wrong since any type of cannon up to whole culverin could be drake-bored. The drake was probably a taper-bored gun, wider at the muzzle than in the chamber. This type of gun was supposed to have been invented by Prince Maurice of Nassau and was restricted until 1630 to the royal service. (*C.S.P.D.* 1629-31, pp. 389, 399). Cf. Ordnance memoranda 4 Oct. 1688 (W.O. 47/19A, p. 55), 18 April 1669 (*ibid.* p. 255), 11 Feb. 1674 (W.O. 47/1913, unpaged) for the sending out of designs for mortars.

[3] 16 Oct. 1663, W.O. 47/5, p. 110.

[4] W.O. 47/19A, p. 46.

bardment, were expected to say whether or not a new idea was sound, especially if to grant the inventor's request for a practical trial cost the Crown very little. The only regular use of government stores for experiments was made in order to improve military fireworks. Frequent trials of new mixtures of combustibles intended for fire-balls and fire-ships to harass the enemy were made, some in the presence of the King himself.[1] Other experiments conducted at the suggestion of an inventor were paid for from public funds when on occasion it was possible to appeal to the King's undiscriminating scientific patronage or to arouse the martial interests of Prince Rupert.[2] Once a gun seems to have been sent to Edinburgh to make an experiment there, and in 1705 ordnance were lent to the Royal Society to make some experiments on sound.[3] According to John Collins, the gunners of the Tower, attended by mathematical science in the persons of Lord Brouncker (President of the Royal Society), Sir Jonas Moore (Surveyor of the Ordnance) and Robert Hooke (curator of experiments to the Royal Society), made trials on Blackheath, then a favourite scene for martial exercises, of the ballistical theories of Robert Anderson, 'the knowing weaver'.[4]

Another active experimenter was Captain Richard Leake, Master Gunner of England from May 1677 to his death in 1696. In this post

> he had full scope for his genius, and he had indeed a surprising one in all manner of fiery productions, so as to excel all the engineers of his time. He had frequent trials of skill with French and Dutch gunners in the warren at Woolwich [where the first ordnance laboratory was set up] at which King Charles and the Duke of York were often present, and he never failed to baffle all his competitors.[5]

He was credited with finding out the way of firing the fuse of a mortar-bomb by the blast of the charge, contrived what he called a 'cushee-piece', apparently a sort of howitzer, and was

[1] E.g. W.O. 47/10, p. 70.

[2] E.g. 11 Nov. 1664 (W.O. 47/6, pp. 64, 133), 14 May 1691 (W.O. 47/10, p. 48), 23 March 1668/9 (W.O. 47/19A, p. 223).

[3] 14 May 1681. *C.S.P.D. 1680–1*, p. 278.

[4] Below p. 125. This is the only occasion of the kind which I have observed, and it does not appear to be noticed in the Ordnance minutes.

[5] Eighteenth-century *Life of Sir John Leake* (N.R.S. 1918), p. 9; W.O. 47/9; Pepys did not share this high opinion of Capt. Leake's ability. *Naval Minutes* (N.R.S. 1925), p. 315.

chiefly responsible for the preparation of the infernal machines directed against St Malo in 1693.[1] It is obvious that there was more encouragement for inventors in the last quarter of the seventeenth century than ever before (not forgetting, however, that the patent system dates from 1617, and that Charles I maintained an ordnance research workshop at Vauxhall), and that this was to no small extent due to the benevolent interest of the royal family in naval and military affairs.[2] Nor does it seem implausible to suggest a relation between this and the scientific, experimental spirit which at times became frankly materialist and mechanical, if it is recognised that scientific inventiveness, countenanced by the Royal Society, and military ingenuity, somewhat obscurely supported through the Board of Ordnance, following separate and not always parallel paths, certainly did not stand in the relation of cause and effect, even though both were to some degree dependent on the same royal patronage.

The Ordnance Board was also used by the Crown in the later years of this period as a means of providing for the maintenance and training of officers to fill the most technical posts in the military service. The duties of these engineers included not only the construction of fortifications but their destruction also, through the technical direction of operations where ordnance, mortars, fireworks and mines were required on an extensive scale. Earlier sovereigns had from necessity employed foreigners in this more technical or scientific branch of the military service, and Sir Jonas Moore in the reign of Charles II was the first Englishman to attain the same standing as the French and Dutch experts. Besides being Surveyor of the Ordnance for many years he was a Fellow of the Royal Society, a writer on mathematics, fortification and gunnery, the friend of Robert Hooke and the patron of John Flamstead, the first Astronomer Royal. In 1685 official action was taken to provide suitable officers in a royal warrant directing that 'divers of our subjects should be well educated and instructed in the art of an engineer and thereby fitted for our service in our fortifications or elsewhere'. Several promising young men who profited

[1] *Life of Sir John Leake*, pp. 10, 17, 23.

[2] Rhys Jenkins, *Collected Papers* (Cambridge, 1936), W. H. Thorpe in *Trans. N.S.* vol. XIII (1932–3).

from the royal bounty, exercised through the board of Ordnance, served with distinction in the wars against Louis XIV. By the end of the seventeenth century the more technical functions in the Army were becoming restricted to an almost autonomous corps of which all the officers had received some training in mathematics and mechanics.

It was to be possible for a long time for titled incompetence to misuse positions of command, but technically qualified officers became to an increasing extent the backbone of the middle ranks of both the Army and the Navy. Military technique no longer suffered because of a lack of scientific principles whereon to proceed; on the contrary, the scientific theory of the arts of war —especially gunnery—was by the mid-eighteenth century far in advance of the practice of the day. It was backwardness in engineering and manufacturing methods which caused the prevailing stagnation in military tactics. While there can be no comparison between the theoretical background in ballistics and artillery science inherited from the seventeenth-century renaissance by one of Napoleon's artillery officers, and the meagre rules-of-thumb which were the sole resource of a sixteenth-century cannoneer, each would at once have been familiar with the *materiel* of the other. In consequence much of the 'science of war' formulated in the late eighteenth century was utterly useless.[1]

As centres of research and training, as influences working towards the technical improvement of artillery and gunnery, the military departments of all the principal states were poor makeshifts. In France Colbert, controlling great wealth and power, and able to attract to the service of Louis XIV many of the finest gun-founders of other countries—Antoine de Chaligny from Lorraine, the brothers Keller from Zurich, Alberghetti from Venice, Besch from Sweden—was also able to found the best artillery school in Europe at Douai, but both gun-founding and military education continued on conventional lines. If France achieved a slight technical superiority over her enemies, it was certainly not proved to be decisive in any of the wars of Louis XIV, and the ballistical studies of the Académie des Sciences were of no practical value. Although Charles II's Ordnance officers had

[1] This point is well brought out by Tolstoy in *War and Peace*.

close contacts with the leading gun-founders, although one of their number was a notable mathematician and F.R.S., neither the wealden industry nor military training benefited from an application of scientific principles. England's special military advantages, her iron ordnance industry, her Navy, her incomparable naval gunners, owed nothing to the patronage of science by the state, indeed, the Royal Society protested that navigation was not its business.[1] About 1650 it would have been possible for an active military department in England to combine, with a certain amount of scientific empiricism, the metallurgical skill of Sweden and Germany, the administrative skill of France, the mathematical skill of Italy, the manufacturing skill of the Netherlands, to reach a new level of warlike efficiency. It did none of these things, but left the gun-founder and the gunner to follow familiar, traditional paths.

4. *The English Ordnance Industry and the State*

The Board of Ordnance in England, like similar departments in other states, was the channel through which the every increasing demands for military supply, in a century of almost unceasing warfare, pressed upon industrial resources. Throughout the seventeenth century the Royal Navy increased in number of ships, and these ships carried more and heavier guns, both of brass and iron, while the requirements of merchant shipping, of forts and garrisons, and of the field armies grew in proportion. From the beginning of Elizabeth's reign there was a steady flow of ordnance from the Weald up the Thames to Tower wharf, control of which contributed materially to the victory of the Puritan revolutionaries. After the restoration of the monarchy the founders were less prosperous, but there was no long period of settled peace, and the expansion of the Navy was continued by Pepys. Single contracts for as many as five hundred new guns were not uncommon, and a single order for 1,500 cannon, weighing in all 2,552 tons, was signed in 1653. Moreover the mortar with fire- or explosive-bomb was more widely employed both on land and at sea for the bombardment of fortified places, and had its companion in the cast iron hand-grenade. Where the ordnance

[1] G. N. Clark, *Science and Social Welfare*, p. 16.

account-books reveal the fireworker calling for such strange materials as antimony, beeswax, brandy, spirits of wine and oil of bitter almonds for his compositions, and the bombardier being furnished with quadrants, mathematical and surveying instruments for the management of his mortars, they show the impact of chemical and mathematical science upon the traditional art of war.

Gun-founding was perhaps the only one of the many industries contributing to the maintenance of an army or a fleet which could not have existed without the market offered by the nation-state, at least in England; Englishmen boasted of their skill in this art, and were willing to make sacrifices to keep it in being. To Sir Walter Ralegh it was 'a jewel of great value, far more than it is accounted by reason that no other country could ever attain unto it although they have assayed it with great charge'.[1] And if the iron gun, cheap and durable, was not suitable for the finest of the King's ships, the English gun-founder could make as good a brass gun as any continental arsenal, too.[2]

To thoughtful men the strategic value of the English iron ordnance industry was a matter of political importance. Since England was not altogether free from dependence upon foreign supplies of small-arms, horse-furniture, powder and armour in the time of the first two Stuarts, it was all the more important to preserve her monopoly of iron ordnance.[3] The Tudors had brought over skilled workmen from Italy and France to lay the foundations of the English armaments industry; now it was necessary to take precautions lest the native skill be beguiled away by foreigners. Since the government was particularly sensitive

[1] *Observations touching Trade and Commerce as it was presented to King James* (London, 1653), p. 35. Cf. Fuller's *Worthies* (1662), *s.v.* 'Sussex', p. 99; Robert Norton, *The Gunner* (London, 1628), pp. 67-8, and many other passages of contemporary comment.

[2] Brass guns were not issued to the garrisons if iron were available, but were restricted to the Navy. *C.S.P.D. 1650*, p. 378; *1651-2*, p. 64; *1661-2*, p. 431. *Acts of the Privy Council, 1625-6*, p. 237.

[3] In 1620, when it was proposed to send an army to Germany, a commission reported to James I that having called before them Mr Edward Evelyn the powder maker, the London gunsmiths and armourers, they found the trade able to supply so little 'that the greatest part of powder and arms for horse and foot must be provided in the Low Countries, where (we conceive) it may best be had' (Francis Grose, *Military Antiquities* [1786], vol. I, p. 363). In Charles I's reign the average delivery of powder into the Tower was only ten lasts per month. When the Scottish war began in 1639 the Crown was forced to buy arms in the Low Countries, and to borrow cannon and ball from the East India Company (Ordnance Records, 1636-9).

to pressure from the ironfounders—who constantly complained that they would be forced out of business if this privilege or that was not granted to them—they took active steps to prevent workmen being seduced overseas by offers of high wages.[1] The government was also under constant pressure from the iron-founders to permit the free export of iron ordnance. To the politician this seemed like placing weapons in the enemy's hands, and since Queen Elizabeth's reign the export of ordnance except to friendly Protestant powers under strict licence had been prohibited with the approval of all who echoed the opinion of Sir William Monson that there was 'no country from the hithermost parts to the uttermost bounds of the world but is . . . stuffed and fortified with them [ordnance], to the unspeakable hazard and danger to ourselves', and opposed the gun-founders' disforesting of Kent and Sussex.[2] The gun-founders, eager to sell as great a number of cannon as possible, whether at home or abroad, were also troubled at the rise of the Swedish iron industry, which a little later provided Gustavus Adolphus with his artillery.

According to a report submitted to the Secretary of State, the Swedish trade began about 1618, guns being brought through Germany to Bremen and thence to Amsterdam, where they were sold at 15 or 16 guilders the hundredweight. The workmen were Germans and Walloons, and though their pieces at first broke in the proof they were now scarcely to be distinguished from those cast in England. If, the writer suggested, 'English pieces were suffered to be carried over, all their furnaces would be suppressed in short time, if not, pieces they must have. And daily they will be more and more experienced in casting of pieces.'[3]

The arguments of the gun-founders against conventional mercantilist economic theory would have delighted Adam Smith. If, they maintained, the export of English ordnance is forbidden, the foreigner will be none the worse since he can buy them in Sweden; in fact the prohibition encourages the Swedish iron

[1] E.g. *C.S.P.D. 1623-5*, p. 489, *Acts of the Privy Council, Jan.-Aug. 1627*, pp. 368, 379.

[2] *Naval Tracts* (N.R.S. 1913), vol. IV, p. 44. He had found English ordnance sold at home for £12 a ton, in Amsterdam for £40, in France for £60 and in Spain for £80 all in one year. Ernest Straker, *Wealden Iron* (London, 1931), pp. 149 *et seq.*, R. H. Tawney and E. E. Power, *Tudor Economic Documents* (London, 1924), vol. I, pp. 231, 262.

[3] *S.P.D. James I*, p. 155, no. 11. Cf. *Acts of the Privy Council, 1623-5*, pp. 136-7.

industry, while English furnaces must be allowed to cool for lack of a market, and the King loses the advantage of a ready supply of guns in time of war.[1] Yet the Monopolies Parliament of 1621, swayed by patriotic economics, exempted the gunpowder and ordnance monopolies from the general abolition contained in its abortive statute.[2] During the 'eleven years tyranny' of Charles I export was permitted—in return for suitably heavy licence fees. During the twenty years after 1640 the ordnance industry had no cause to look for foreign markets which by the end of the Interregnum had largely been lost to Sweden. In 1667 the ironfounders were reduced to petitioning against the importation of foreign iron, though their industry still prospered in time of war.[3]

Thus the rise of the Swedish iron industry finally solved what had formerly been a political dilemma: to allow export, encourage the gun-founders but outrage public opinion, or to forbid export, tolerate a monopoly, and pay high prices for every gun admitted to the Tower. The Crown never considered, apparently, the *étatisme* developed by Colbert in France, where the manufacture of weapons of war was restricted to royal factories or large private establishments controlled by government supervision, and where the small master was driven from all but the private trade in luxury weapons for the chase.[4] In England the ordnance industry remained in a position that was neither wholly private nor wholly public, receiving just sufficient encouragement from the Crown to enable it to exist. Thus Elizabeth had created the title of Royal Gun-founder, which was held by various members of the Browne family (controlling many furnaces in Kent and Sussex) throughout the greater part of the century, but the practical business arrangements of the royal gun-founders were far more important to them than their title or their fee of four shillings a day.

[1] *C.S.P.D. 1633–4*, pp. 358–9.

[2] The debates are reported in Notestein, *Commons Debates of 1621*, vols. IV, VII.

[3] According to this petition there were thirty-four furnaces blowing in Kent and Sussex during 1653 (First Dutch War), of which only fourteen remained permanently in use. When war broke out again in 1664 eleven more were relit and 'stocked upon the war and future encouragement'. In the Second Dutch War at least eighteen furnaces cast guns and shot, but the other nine, ruined after 1653, were not restored (*S.A.C.* vol. XXXII, pp. 21 *et seq.*). Some of Prince Rupert's annealed and turned guns were exported to France in the 1680's (*C.S.P.D. 1680–1*, pp. 493, 559, 594–5; *1682*, pp. 8, 10, 129).

[4] Boissonade, *op. cit.* pp. 93, 95.

Thus John Browne, who succeeded his father as Royal Gunfounder in 1618, profited directly and indirectly from the brisk foreign policy of Buckingham and the poverty of Charles I. He co-operated with the Navy in experiments to make iron guns less weighty, for which he received a special reward of £200.[1] He commenced casting in brass as well as in iron. With the Long Parliament and its Committee for Ordnance he made many satisfactory business agreements, always advocating the advantages of a 'certain, full expeditious supply with good and serviceable commodities' only obtainable from a firm protected by monopoly; as he did later with Cromwell's officers of the Ordnance in the course of the Dutch and Spanish Wars. John Browne died in 1651, but the management of his firm and the royal title remained with his sons, George and John, until 1681. After the death of John II, however, the firm fell into great difficulties, because the administration of Charles II was reluctant to pay debts, and the manufacture of Prince Rupert's guns, which had been begun, proved to be an unprofitable venture. Its collapse was part of a general business depression which began in 1682.[2]

John Browne was no mere tradesman, indeed he was described by a rival as 'noe workman. Hee never served to the Trade nor can he sett his hand to the manufacturie and workmanship of a Peece', but he controlled from his house in St Martin's Lane a partnership which was probably one of the largest commercial organisations in seventeenth-century England. After his death a formal partnership agreement was drawn up between the iron-founders who co-operated with the Brownes in supplying the Board of Ordnance, including—besides George and John Browne, with their furnaces at Buckland in Surrey, Spelmonden and Horsmonden in Kent—Alexander Courthope, who owned furnaces at Horsmonden, Ashburnham and Bedgebury; William Dyke, with an ironworks at Frant in Sussex; Thomas Foley, the nail-maker of London; and several other gentlemen controlling furnaces in different parts of Surrey, Sussex and Kent. They

[1] *C.S.P.D. 1625–6*, pp. 171–3, 280, 320; *1615–49*, p. 74; *Addenda, 1580–1625*, p. 639; Notestein, *op. cit.* vol. VII, p. 419; *H.M.C. Earl Cowper*, vol. I, pp. 116. 126; *13th Report Appendix*, pt. IV, p. 178; *C.S.P.D. 1631–3*, p. 499.

[2] *C.S.P.D. 1660–1*, pp. 14, 281, 283, 385; *1677–8*, p. 451; *1680–1*, p. 559; G. N. Clark, *The Later Stuarts* (Oxford, 1934), p. 104.

appointed their own attorney to do business for them in London, who also acted as banker.[1] George Browne was the business head of the company, as Royal Gun-founder negotiated the contracts with the Board of Ordnance, and decided when it was necessary to re-commence casting at each furnace, but probably Alexander Courthope, a wealthy Sussex squire, had the greatest share in the actual casting, for he owned the two largest blast-furnaces in England.[2] It seems likely that this organisation constituted a virtual iron-trust, controlling every important foundry in the Weald, on which the Crown was entirely dependent for its supply of artillery.[3] Later many people (including the Drapers' Company and some frequenters of the court of Charles II) who had no connection with the industry invested money heavily in the Browne Copartnership. In 1685 it was reckoned that the sums owing from the Crown (with interest) amounted to £43,000, of which some £23,000 was due to a dozen principal investors and the original patentees of annealed iron ordnance, and the remainder to forty-five lesser creditors.

In this last section I have shown that ordnance manufacture was no Cinderella among English industries. It exercised the thoughts of politicians almost as much as the cloth industry or shipbuilding; it held first place, with navigation, in the thoughts of those who were concerned for the development of the island's imperial might. But it should not be supposed that the ironfounders were able to hold the Crown to ransom, or extract any concession they sought by reason of their strong bargaining position. The Ordnance Board could always import from abroad, and of course the prosperity of the industry—since it was not popular among the gentry or other trades, being a great consumer of timber—depended entirely on the toleration of the government and merchant shipowners. It would have collapsed if forced to

[1] Various documents referring to the 'Co-partnership . . . in and about the making and casting of Brass and Iron Ordnance, shot, and other things' are printed in *S.A.C.* vol. XXXII, pp. 26 *et seq.* or are available in the War Office Records (W.O. 47/7, p. 77 r.; 19/A, p. 425) and the *Courthope Papers*. The furnaces were sometimes described collectively as the property of George Browne, which they were not.

[2] For instance, 190 of 400 guns cast in 1667 came from Horsmonden and Ashburnham.

[3] E.g. George Browne writes to Courthope (24 Dec. 1664): 'We must of necessity set Bedgebury to work on Guns for they [i.e. the Ordnance Officers] are in such want yt they will have all our Di-Cull. of what length so ever & ask for more than we are able to make of all sorts and sizes' (*Courthope Papers*).

rely on the domestic trade in firebacks, kitchen pots and grave slabs. Even the railing of St Paul's Cathedral was nothing compared to the ordnance trade, and the price of iron, owing to competition from Sweden and the Forest of Dean, tended to fall rather than rise throughout the century.

We are here more concerned, however, with the scientific than the economic history of the English ordnance industry. The point I wish to make is that its influence on science, or of science upon it, was negligible. The Officers of the Ordnance were not influenced to any great extent by the progress of the scientific revolution, and the Board itself exercised but slight influence on the techniques of ordnance manufacture. Administrators, like scientists were only interested in the uses of machines; their manufacture was the proper province of the craftsman. We do not find that there was in practice that interworking of science and industry which some members of the Royal Society—and its first historian—believed to be desirable, or that the machinery of the state was suitable to bring this about. Indeed, the attempt to do so was scarcely made. The relations of the Board and the Browne Copartnership were purely business relations, the Crown was a consumer, and had as little influence on the design of the manufactured articles it bought as any other consumer. These relations have only one guiding thread, a political one; the desire to keep the Crown strong and well-defended; to buy cheaply; to satisfy the country gentlemen who controlled the purse. In this respect the ordnance industry was in no special position. Official policy regarding it was exactly the same as that regarding the cloth trade or agriculture. The idea that science had something to teach the craftsman was a commonplace among its seventeenth-century publicists, but the idea that science might add something to the art of war had not yet affected the traditional procedures of military bureaucracy.

CHAPTER II

THE BACKGROUND OF SCIENCE II THE GUNNER AND HIS ART

From the beginning of the sixteenth century a number of new factors favoured the ever-increasing output of military text-books. First, obviously, was the invention of printing, and second the development of new weapons and techniques. Officers and men had to be trained in the drill of musket or pistol and their tactical combinations with pike and sword; in the deployment of artillery and mortars; in sapping and mining, in bridge-building and the provision of supplies for larger armies which no longer lived exclusively on the country. Skill in war was no longer the natural birthright of every gentleman, it had to be learnt by study and experience. Even the literary renaissance contributed an interest in the disciplined formation and tactical efficiency of the Roman legion, which received the high praise of Machiavelli. If until the time of Cromwell the English writer on military affairs was little more than a translator of Italian, Spanish and French authorities, if the English soldier was unfamiliar with the subtle technicalities of positional warfare on the continental pattern, these deficiencies were more than supplied by the islander's prowess as a fighting seaman and a naval gunner. The esteem in which the sea-gunner was held shines from the reminiscent pages of retired commanders —Ralegh, Monson, Smith, Sturmy and the rest—though Pepys was to write regretfully, of the pre-eminence of former days, 'we have since lost it'.[1] England's success in naval warfare owed less to any advantages of technique or skill than to sheer persistence in battle. Out-manoeuvred by the Dutch, out-sailed by the French in many a campaign, the English commander disdained exchanges of shot at long ranges, relying upon the brutal destruction of the enemy's crew and hull at close quarters to win the day.[2]

[1] *Tangier Papers* (N.R.S. 1935), p. 419; *Naval Minutes*, p. 315.

[2] William Bourne, *The Arte of Shooting in Great Ordnaunce* (London, 1587) writes: '... we Englishmen have not been counted but of late days to become good gunners, and

That Englishmen could take pride in their naval gunnery was rather owing to the surviving tradition of Elizabethan seamanship than to any active participation of the Crown in the selection and training of ship's gunners. Master-gunners were indeed appointed directly by the Admiralty and candidates for appointment were supposed to be examined by the Master Gunner of England, but the discipline of this official scarcely extended beyond the established or fee'd gunners of the Tower, and his facilities for training, in the old Artillery Garden (near the present Liverpool Street station), were very limited. Administrators paid more attention to the conservation of stores than to a finished drill and half-a-dozen preliminary rounds at sea were judged to be sufficient before commencing a campaign.[1] Yet Master-gunners, we are told, 'especially those of great ships, wore their swords on shore, kept company with the Commission officers, and were much respected by all,' for the post sometimes held the beginnings of a noble career.[2]

The obverse of complacent satisfaction in the supremacy of English gunnery was fear of a lack of just such trained experts. This was particularly acute in the period before the Civil War; afterwards expressions of concern became less frequent.[3] Yet as late as 1669 Captain Samuel Sturmy wrote that he was

ashamed to hear how senselessly many sea gunners will talk of their art, and know little or nothing therein, but only how to sponge, lade, and fire a gun at random without any rules of finding the dispart,

the principal point that hath caused Englishmen to be counted good gunners hath been for that they are hardy or without fear about their ordnance, but for the knowledge in it other nations and countries have tasted better thereof' (Preface to the Reader, sig Aiii).

[1] See the Duke of York's orders for Master-gunners 1663 (P.R.O.SP. 29/72 no. 13).

[2] *Life of Sir John Leake*, vol. 1, p. 2.

[3] 'What a number there be that will take upon them to be gunners, yea, and that master gunners, that are not sufficient nor capable in these causes, but are in that respect altogether ignorant, standing upon no other thing but their antiquity, that they have served as gunners so long time' (Bourne, *op. cit.* Preface). In 1637 the Lords of Admiralty wrote to Trinity House and the Master of the Ordnance for names of gunners fit to be master-gunners of ships (*C.S.P.D. 1637–8*, p. 13). Ordnance Minute, 25 Jan. 1695: 'That a letter be writ to Capt. Leake to make out a list of all the fee'd gunners and Practitioner gunners showing how they are disposed of, to the end it may be seen what of them may be drawn out for future service' (W.O. 47/18, p. 225). The ballistics 'expert' had of course an axe to grind in exposing the theoretical ignorance of gunners, but official records agree in showing that there were few men who rose above the rudiments of the art.

thickness of the metal in all places, and proportion any charge of powder thereto, and other rules which should be known.[1]

In the years before the Armada a suggestion for setting up a corporation of gunners and compelling all merchant ship-masters to appoint men as gunners only if they were certified as capable by the Master of the Ordnance had been put forward.[2] This scheme failed and the 'scholars of the Artillery Garden' or 'fee'd gunners' remained the only professional artillerists in the country.

Elizabeth had ordered that they be exercised four times a year, but if William Bourne is to be believed their ignorance was striking and the system of instruction outmoded. The fee of sixpence a day attached to these places ensured that they should become humble sinecures in the patronage of the Ordnance Officers. Attempts were made to correct this corrupt and perilous condition, for instance an order of 1668 that the fee'd gunners should be exercised by the Master Gunner every Wednesday—in fine weather. When Sir Thomas Chicheley was Master of the Ordnance (1670–74) he was commanded to dismiss incompetent men, to put down the sale of gunners' places and to fill vacancies with men trained in the Artillery Garden, who were to be made to conform to proper written instructions.[3] The chaos in English administration to the end of the century contrasts with the organised methods developed in other countries; Spain had established artillery training schools before the end of the sixteenth century, and Catholic gunners even had their own patron, St Barbara. In Richelieu's time artillery training was given to French seamen in the principal ports and dockyards, a system which Louis XIV improved in 1679 by founding a school at Douai, where the course of instruction was set forth by the Grand Master of the Artillery himself.[4]

The English land artillery was not organised regimentally until

[1] *Mariner's Magazine or Sturmy's Mathematical and Practical Arts* (London, 1669), revised 1684, sig. a verso.

[2] *S.P.D. Elizabeth*, p. 147, nos. 94, 95; p. 157, nos. 40, 41, 42. Printed in G. A. Raikes, *History of the Honourable Artillery Company* (London, 1878), vol. I, Appendix D.

[3] W.O. 47/91A, p. 47; *C.S.P.D. Addenda, 1660–85*, pp. 449, n.d.

[4] Luys Collado, *Prattica Manuale dell'Artiglieria* (Milan, 1606, first published in Italian Venice, 1586), pp. 339, 384 *et seq.*; F. B. Artz, 'Les Débuts de l'Education Technique en France, 1500–1700'; *Revue d'Histoire Moderne*, vol. XII (1937), p. 510; Saint-Rémy, *op. cit.* vol. I, p. 39.

1716. Before this each train of artillery required for a military expedition was detailed *ad hoc* by the Board of Ordnance. Such a train would be commanded by a General or Master of the Artillery, under whom the immediate supervision of the gunners and their assistants (called matrosses) rested upon a Master-gunner.

Contemporary military writers dwell on the high qualities—technical and moral—needed for these positions of command, and relate in detail the knowledge to be instilled into the gun-crews, which may be divided into two parts, practical and theoretical. In the first place the practical gunner should know how it 'dispart' his piece, that is, to set up a foresight on the muzzle equal to the difference in the thickness of metal at the muzzle and the breach, so that a line of sight over the base ring touching the tip of the dispart would be parallel to the axis of the bore. Then he should know how to set his under-gunners at their stations about the cannon, and to guard his powder from a chance spark; how to fashion a ladle of the correct size for the charge; how to sponge and load the gun, using the rammer with care to avoid mischances; and finally how to train the gun on the target by eye and carefully fire it from his smouldering linstock, being always wary of the recoil. For the naval gunner Sir William Monson adds that it is a principal thing in him to be a good helmsman, to know when to call to the man at the helm of the ship to luff or to bear up in order to trim her more level, and to know from the heaving of the sea when best to give fire.[1] In judging the right moment the sea-gunner had to make proper allowance for the time that the priming and charge took to burn and for the movement of his own ship and that of the enemy, except in a stern chase. It was necessary to shoot twenty paces in front of a vessel under sail at the extreme useful range of about a thousand paces.[2]

This was the sort of lore which a novice in the art of gunnery might be expected to possess. An ignorant seaman or soldier might know how to load and let off a gun point-blank at the foe, but in the eyes of the military experts the test of the true gunner was his skill in its loftier realms, especially in the mathematical

[1] *Naval Tracts of Sir William Monson* (N.R.S. 1913), vol. IV, p. 33.

[2] Thus Fournier, *op. cit.* (1643), p. 139.

parts where gunnery became a science. A definition of the new science—for which the German writers coined the name of *geometrische Buchsenmeisterei*—was first given by Niccolo Tartaglia, the founder of ballistics and the art of gunnery, in 1537.

There are two principal parts [he wrote] necessary to a real bombardier wishing to shoot by rule ['con Ragione'] and not haphazard, each of which is nothing without the help of the other. The first is that he should know approximately and work out from its situation the distance of the place which he has to batter. The second is that he should know the range ['quantita de tiri'] of his cannon, according to their different elevations; knowing these things he will not err much in his shooting; but lacking one or other of these he cannot on any account shoot by rule but only by discretion, and if by chance he strikes the target, or near it, at the first blow, this is through good luck rather than knowledge ['scientia'] especially in distant shots.[1]

Nor were these ideals altogether unapproachable, for his contemporary Biringuccio speaks of instruments for measuring distances, levelling cannon and so forth as though they were in everyday use.[2] An important section of most of the manuals of practical geometry was devoted to mensuration, frequently with illustrations of the gunner measuring his ranges, heights and depths. One of the instruments usually explained was the 90° quadrant first described, if not invented, by Tartaglia, in the simplest form of which an arm was pushed into the barrel while the angle of elevation was read off against a plumb-line.[3]

In every European tongue primers in artillery were published, commencing with the rudiments of arithmetic and geometry and continuing through simple surveying and the use of proportions to the theory of gunnery. Tartaglia had already pointed out the importance of measuring heights and distances; the method of proportionals and of working out square roots was necessary to calculate charges, compare the weights of cannon balls of different materials and sizes, and to solve the sort of problem occurring in the use of range tables, to find x in the expression

$$375:500 = x : 650.$$

[1] *La Nova Scientia* (Venice, 1537), Epistle Dedicatory, sig. Aii.

[2] *Pirotechnia* (1943), pp. 418 *et seq.*

[3] His priority is disputed by M. Jähns, *Geschichte der Kriegswissenschaften* (1889), Abt. I, pp. 410, 598.

A writer who was so unusual as to ignore these mathematical excursions took the trouble to justify his omission.[1] It is not difficult to realise that the type of knowledge outlined in this paragraph is of a very different quality from that described in the one preceding, for this is the work of the head and that the labour of the hands. In other words, the official outlook and instruction suitable for the manual gunner would not satisfy the theoretician.

The importance of skill in calculating for warlike operations was first exhibited to Englishmen by Leonard Digges in his *Pantometria*, a work completed and published in 1571 by his son Thomas, later muster-master of Leicester's army in the Netherlands. Of this work Thomas wrote in his Preface that the reader would find Stereometry useful in all sorts of building work, Planimetry would serve 'for disposing all manner grounde plattes of Cities, Townes, Fortes, Castles, Pallaces or other edifices' and for drawing out the lines of a camp, while

The other parte named *Longimetra* the ingeniouse practizioner will apply to Topographie, fortification, conducting of mines under the earth, and shooting of greate ordinance. So that as there is no kinde of man of what vocation or degree soever he be, but shall finde matter both to exercize his witte and diversely to pleasure himselfe, so surely for a gentlemen especially that professeth the warres, as well as for discoveries made by sea, as fortification, placing of campes, and conducting of Armies on the lande, how necessarie it is to be able exactly to describe the true plattes, symetrie and proportion of Fortes, campes, townes, and countreys, coastes and harboroughes. . . And for science in great ordinance especially to shoote exactly at Randons (a qualitie not unmeete for a Gentleman) without rules Geometrical, and perfect skill in these mensurations, he shall never know anything. . . .[2]

Although neither of the Digges is now much remembered,[3] *Pantometria* and *Stratioticos* (1579) were still read in the seventeenth century and the simple lesson that mathematics was important because of its certain rules, was not forgotten. Robert Norton in *The Gunner* (1628) wrote of geometry as the

[1] Gabriel Busca, *Instruttione de Bombardieri.*

[2] *A geometrical Treatise named Pantometria divided into three Books* (London, 1571), sigs. Aii, Aiii, 'at Randons', or Randoms=at angles of elevation above the horizontal.

[3] Of both there is an (inaccurate) account in *D.N.B.* Some space is devoted to them by F. R. Johnson in *Astronomical Thought in Renaissance England* (Baltimore, 1937). Digges' scientific work apart from astronomy has never been properly noticed.

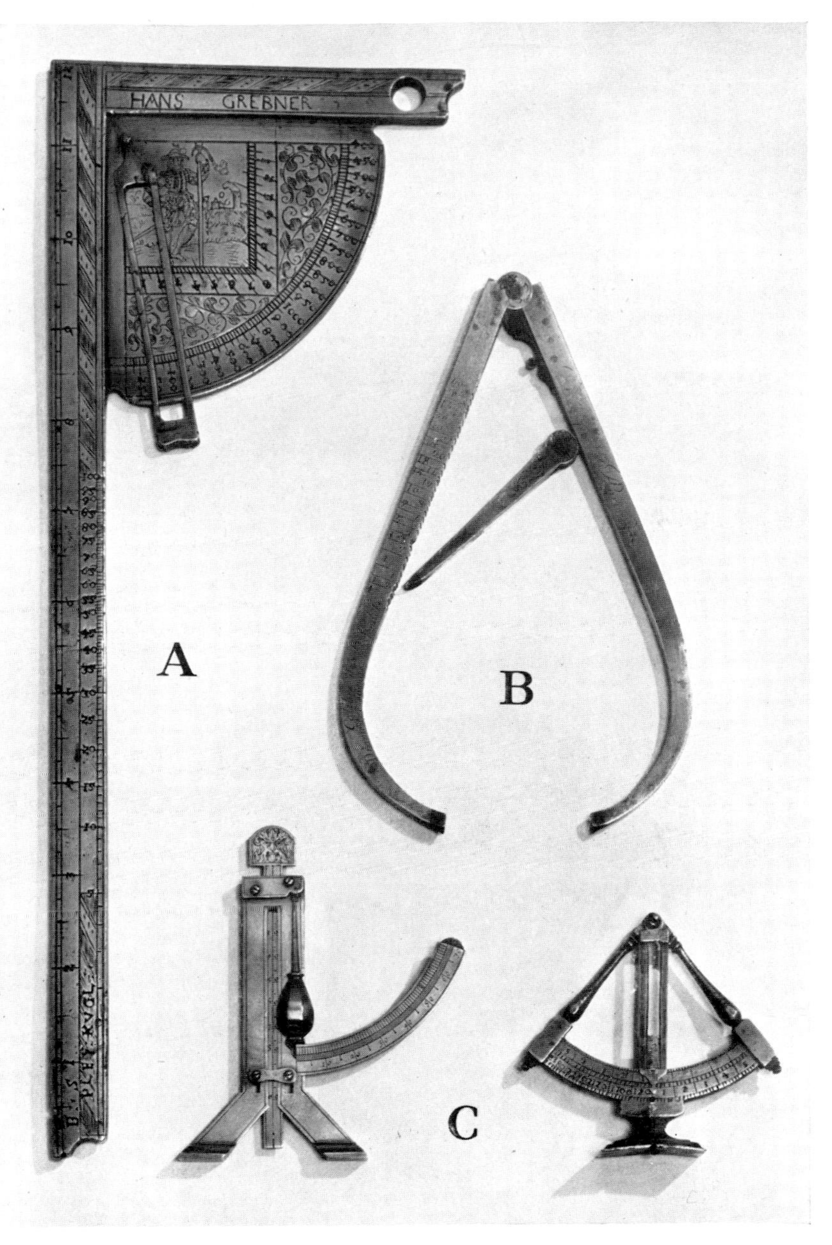

GUNNERS' INSTRUMENTS OF THE SEVENTEENTH CENTURY

A. Quadrant or square (German); B. Calipers (French); C. Clinometers

Art to measure well and the sinews of the Art of Artillery.... For it is necessary that the gunner should know what a line, a superficies, and a body is and how to measure each of them, as well the right as the crooked, the level Hypotenusal and perpendicular and diametrical lines, and the angles they can make right or oblique.

He expected the gunner to understand the use of trigonometrical and logarithmic tables.[1] Samuel Sturmy, author of a self-educator, thought that the gunner

ought to have skill in arithmetic to work any conclusion by the single and double rule of three, to abstract both the square and cube roots, and to be perfect in the Art of decimal arithmetic, and to be skilful in geometry, to the end he may be able through this knowledge in these arts, to measure heights, depths, breadths and lengths and to draw the plat of any piece of ground.[2]

Captain Thomas Binnings reverses the argument, making the usefulness of gunnery arise from its foundations in science, and military success depend upon the mathematical attainments of the combatants.[3]

If in weighing and measuring science was a necessary auxiliary of the gunner's art, in the 'art of shooting great ordnance at randoms', that is, the elevating of cannon above the horizontal to strike targets beyond the point-blank range, the real mystery of gunnery was disclosed and to it science was believed to hold a master-key. It was penetration of this mystery 'scarcely thought on, and farre beyond the compasse of ordinary Cannoniers without exquisite knowledge in the *Mathematical Sciences* to intermeddle withall', requiring a familiarity with the 'several Proportions of the mettaline Bodies and Soules or Cylinders of all several Peeces and of the strange varietie of the Circuites of all Bullets in the Ayre, by reason of the repugnance or variance of the violent and natural motions' which was the sign of the master and the pride of the men of science who despised rule-of-thumb gunnery.[4]

[1] *Op. cit.* pp. 23, 25, 94-5.

[2] *Mariner's Magazine*, bk. v, p. 45.

[3] *A Light to the Art of Gunnery* (London, 1675). To the Reader. Cf. Capt. John Smith, *Seaman's Grammar* (ed. of 1692), p. 91; the gunner should be proficient in 'Arithmetic both Vulgar and Decimal; whereby he may be able to work the Rule of three (or Golden Rule) both Direct and Reverse, to extract the Square and Cube Roots &c.; in Geometry, whereby he may be able to take Heights, Depths and Distances; to take the true Plat of any piece of Ground; and thereby to Mine or Countermine under the same, or any part thereof.'

[4] *Pantometria* (1591), p. 175.

The gunnery experts of the renaissance age were attempting a union of natural philosophy and mathematics, and in this respect it is only fair to give them their due place in the scientific revolution. They had many faults. They were uncritical, they mouthed without meaning the Aristotelian vocabulary of science; they stated that things were found from experience which palpably were not always so found; yet long before the time of Galileo they were trying to submit philosophy to the yardstick of computation, had yielded to the magic of numbers and realised with practical good sense that a mechanical theory is vain speculation unless it can be checked by measurable results. Gunners were weighing and measuring charges, angles and distances before measure became the great instrument of physical science, in fact we can see in their activities a conspicuous example of the truth that accuracy with balance and ruler will prove very inadequate without the framework of an intellectual system. This Galileo provided, for whereas the practical men had failed to make a reconciliation of traditional philosophy and empiricism, he after examining the physical principles of the science of the schools and finding them incapable of mathematical expression, rejected them and looked for new laws.

The founder of the theory of gunnery which was expounded into the second half of the seventeenth century was Niccolo Tartaglia, himself a victim of French military predominance, having received a disfiguring wound during the sack of Brescia (1512). Tartaglia's theory of motion and the philosophy from which it derived will be touched upon in a later chapter; for the present the more practical part of his writings, always admitted to be of great authority even by those who criticised them, are of interest.

Two factors contributed to the initial success of his ideas: Tartaglia was a theorist without experience of artillery except that which he gained in talk, and he did not set out to differ seriously from the philosophy of the schools.[1] It was not necessary for those who followed him to break with any accepted ideas;

[1] See the Epistle Dedicatory to *La Nova Scientia*, *iii r. In fact his exposition of the impetus theory is fundamentally opposed to the dynamics of Aristotle. But this was a dispute over explanations, for these two schools of physics were agreed on observation, on which only the Galilean differed.

on the contrary, ignorant of the real incompatibility between the peripatetic system and the theory of impetus, the humbler of them believed themselves to be safely in the centre of the intellectual stream. Tartaglia, though in some respects his new science was even more novel than Galileo's (for no one had written on ballistics before), had not to overcome those obstacles of prejudice and misunderstanding (even among the enlightened) which confronted Galileo.

Tartaglia relates that his interest was first drawn to the theory of the motion of projectiles in 1531, when he was asked at what angle of elevation a gun would shoot the furthest distance; the first-fruit of his studies in this field new to any mathematician was *La Nova Scientia*, published in 1537.[1] This is a small quarto of 78 unnumbered pages, divided into three books, of which the third is devoted to mensuration. The subject of Book I is the theory of the motion of heavy bodies ('corpi egualmente gravi') which are defined as those bodies whose movements are not sensibly affected by air resistance.[2] Book II discusses the shape of the trajectory of projectiles. The whole of Tartaglia's doctrine really derives from the opposite characteristics of violent and natural motions, the former being more swift as it approaches its end, the latter at its commencement.[3] These are both 'common sense' principles, one learns from everyday experience that the longer a fall, the heavier the impact on striking the ground, while to throw with effect it is necessary to be close to the target. From this starting-point Tartaglia was able to show that when a body moves its motion must be either purely violent or purely natural, for if it were not so, the body must move both faster and slower at the same time which is absurd (Prop. V). At the end of Book I, then, the position which Tartaglia has reached in his argument

[1] *iii r.

[2] 'Corpo egualmente grave e detto quello che secondo la gravita della materia & la figura di quella e atto a non patire sensibilmente la opposition di l'aere in alcun suo moto' (bk. I, Definition I).

[3] Tartaglia's definitions of natural and violent motion are as follows: (Definition VI) 'Movimento naturale di corpi egualmente grave e quello che naturalmente fanno da un luocho superiore a un altro inferiore perpendicularmente senza violenza alcuna.' (Definition VII) 'Movimento violente di corpi egualmente gravi e quello che fanno sforzamente di giuso in suso, di suso in giuso, di qua & di la per causa di alcuna possanza movente.' Any motion unless directly towards the centre of the earth is therefore violent (Propositions I and III).

is that a heavy body when projected by a force ('possanza movente') moves for a space in violent motion gradually diminishing in velocity, and then falls into a natural motion, its least velocity and effect being at the point where the transition takes place. He depicts this in a figure showing the path of the projectile as initially a straight line at an oblique angle to the horizon connected by an arc to a vertical straight line leading towards the centre of the earth. The point of least velocity is not at the vertex of the curved segment but at its end.

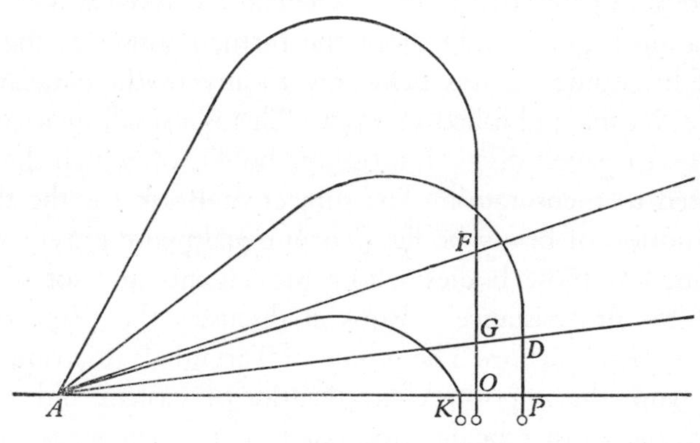

Figure I

It remained for Tartaglia to elucidate this trajectory further and to draw what conclusions he could from it. Accordingly in Supposition II of Book II he declared that any trajectory of a body in violent motion, unless it were perpendicular to the horizon, consisted of rectilinear and curved portions, the latter being an arc of circle.[1] He qualified this by acknowledging that in an oblique violent motion no part of the path can be a geometrically straight line because the weight of the body continually draws it down towards the centre of the earth. 'None the less that portion which is insensibly curved I shall suppose straight and that which is evidently curved I shall suppose part of the circumference of a circle because thus I shall neglect no important matter.' The projected body moves in this arc of circle until at a certain point its motion is perpendicular to the earth, after which it moves

[1] 'Ogni transito over moto violente de corpi egualmente gravi che sia fuora dalla perpendicolar de l'orizonte sempre fare in parte retto e in parte curvo fara parte d'una circonferentia di cerchio.'

along the tangent which is, of course, the path of natural descent. The most distant point which any projectile can attain on any plane is that where the line of vertical descent meets the plane (in the figure on page 38 the points *F*, *G*, *O* on the highest curve, *D*, *P* on the middle curve and *K* on the lowest).[1]

From this reasoning it followed very simply that in the case of horizontal projection the curved segment of the trajectory is a quarter of the circumference of a circle, that in oblique projection above the horizon it is greater than a quadrant, and below

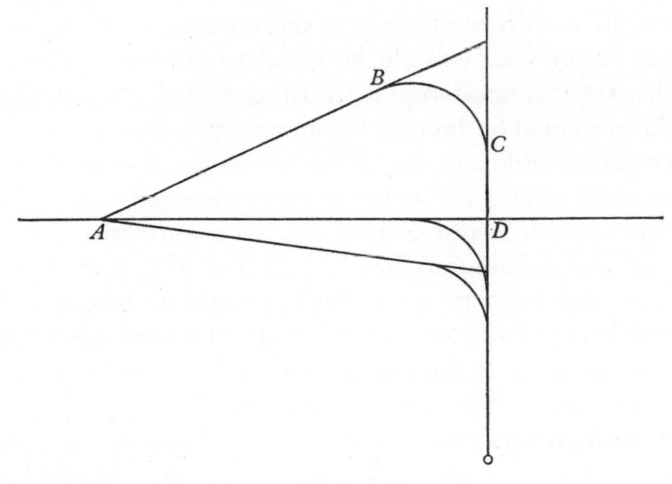

Figure 2

the horizon less.[2] Having established the fundamental nature of the trajectory Tartaglia went on to point out the deductions which might be useful in gunnery. Proposition VII proves that, all things being equal, all trajectories made at the same elevation above the horizon are similar and proportional; that is, for any trajectory *ABCD* as long as the angle *BAD* is constant the ratios of the lines *AB*, *BC*, *AD*, to each other are constant, so that the range *AD* is proportional to *AB*, which is proportional to the strength of the moving force or, as we should say, to the velocity of the projectile. Proposition VIII shows that the greatest range is obtained when the angle of projection is 45°, because the range increasing from a minimum at 0° to a minimum at 90°

[1] Suppositions III and IV.

[2] Propositions IV, V and VI.

through a maximum, it is obvious that the maximum must occur between the horizontal and the vertical. This deduction Tartaglia claimed to have verified by experiment in 1532, when for a wager a culverin of twenty-pound ball was fired with equal charges at 30° and 45°, the range at the first elevation being 11,232 (Veronese) feet, at the second 11,832 feet.[1] Finally (Prop. IX) he asserted that the rectilinear violent motion of a trajectory described by a projectile fired at 45° to the horizon was equal to four times the point blank range of the shot fired with the same force.

Tartaglia also drew attention to various other properties of his ballistic theory which might be useful to gunners, such as the fact that the extreme range is ten times the point-blank range; that the gun could be directed to shoot a certain distance by either of two paths at different elevations, the means of doing so being 'cose non piu audite ne d 'alcun altro antico ne moderno cogitato'. He claimed that he had arrived at a theory for the increasing range of guns and mortars from degree to degree, so that proper angle for shooting any given distance could be found at once from the known range at any other angle.[2] These inventions were never made public, as the fourth and fifth books of the work for which they were destined never appeared.

In 1546, however, Tartaglia brought out a second volume under the title *Quesiti et Inventioni Diverse*, of which the first two books continue the discussion of ballistics.[3] In this work he gave up the didactic form for dialogue, sometimes quite spirited in its style, because his purpose was to assemble his thoughts on several scattered topics relating to the use of artillery rather than to compose another coherent treatise on philosophical and mathematical principles. He gave a more detailed description of the gunner's quadrant. He described the increase and decrease of range as the gun is elevated, remarking in his account that the differences in ranges were least when the ranges themselves were greatest. From their properties he intended (as he promised in 1537) to construct a range table, but again the task was unfulfilled. From the fact discovered in *La Nova Scientia* that the rectilinear

[1] Epistle Dedicatory, *iv verso.
[2] Epistle Dedicatory, sig. Ai verso.
[3] Venice, 1546. The whole treatise is in nine books.

violent motion was four times as long as the point-blank he deduced the proof of the proposition that if it was necessary to bombard a castle on a hill it was better to shoot from below than from an equal height.

Once again he declared, and more strongly, that no part of the trajectory is a straight line: 'non solamente la [colabrina] non tirara li detti passa 50 per linea perfettamente retta ma non tirara un passo solo'.[1] To explain this, which was very contrary to the common opinion, he asked the question, why does the ball, since it moves in a straight line at first, cease to do so? The answer must be, because the great velocity of the ball at the moment when it is shot forth from the mouth of the gun is very much reduced after a short space of time, and as its velocity decreases, so the ball tends to fall to the ground. But, argued Tartaglia, if the velocity of the cannon-ball continually decreased from the very moment it left the barrel, it would be necessary to conclude that the portion of the rectilinear trajectory nearest to the mouth of the gun was straighter than that more removed, which is absurd. The only alternative was to suppose it to be curved from the mouth of the gun. Similarly with the aid of diagrams he was able to clear up various problems occurring in the sighting of guns at point-blank ranges owing to the intersection of the line of metal of the cannon with the axis of the bore.[2]

What is the basis of common sense in Tartaglia's ballistics? He was a theorist indeed, but one who accepted most of the received opinions of the experts in artillery and rarely suggested that they must be mistaken in the observations they reported; rather he undertook to give the philosophical and mathematical explanations. Whether a story was true or false he was not always able to distinguish and therefore was led into rationalising many tales which seemed incredible even in a credulous age. His demonstration that the projectile never moves along a perfectly straight line seemed revolutionary, but in defining the trajectory as rectilinear for all useful purposes for the first few yards of point-blank range he was at once in accord with universal belief and within the limits of accuracy of artillery. The theory that an arrow shot into the air falls vertically downwards was confirmed

[1] *Op. cit.* 11 r.

[2] *Ibid.* 16-17.

by the authority of Leonardo and no one had yet dared to put it in question.[1] That it is possible to shoot further in a straight line if the aim is inclined upwards rather than horizontal was easy to believe—because horizontal motion is 'most curved' and vertical motion 'least curved'—and is not without truth, since the rate of curvature of the trajectory is greatest about the vertex. His doctrine of proportionality between point-blank and extreme range was confirmed by the experience that both are small in the case of short-ranging guns throwing light shot and both long when a high-velocity, heavy-shotted gun is used. In later times this became a maxim of those artillerists who advised the use of powerful weapons, not because of their greater range and penetration, but because they shot straight for longer distances and avoided the uncertainties of 'random' shooting.

Tartaglia's picture of the trajectory, which may well seem to be the grossest of his mistakes since he embellished it with the formal certainty of geometrical demonstration, has the excuse of being an attempt to follow the scientific or Galilean plan of leaving out the small, irrelevant minutiae of observation in order to reduce physical observations to their simplest forms. Taken broadly, Tartaglia's drawing was not such a bad guess, and the rough outline of his figure is nearer to Newton's hyperbola than to Galileo's parabola. But the longevity of his ballistics is obviously to be attributed mainly to his success in agreeing with the experience of professional artillerists, even to their prejudices and absurdities, and in this respect Tartaglia's theory was superior to the parabolic hypothesis of Galileo, which every practical test showed to be at fault.[2] If the latter had the advantage of theoretical perfection, Tartaglian ballistics were adaptable to the satisfaction of the empiricist.

The translation of his works, the flattery of his imitators and plagiarists, are sufficient proof that Tartaglia's writings attained popularity, and there were few further contributions to the theory of gunnery before the middle of the seventeenth century, if one

[1] Leonardo MS. M (Bibliothèque de l'Institut de France, ed. Ch. Ravaisson-Mollien, Paris, 1889—), 53 r.

[2] Partly because the mathematically precise Galilean mechanics could be put to tests of an order of accuracy far greater than was possible with any previous physical theory, so that factors like air resistance at once become significant.

excepts a certain pseudo-Aristotelian elaboration of the type ridiculed by Galileo.[1]

Some novel notions were set out by Thomas Digges, who promised but never published a work on pyrotechnics and great artillery. However, at the end of the 1579 edition of *Stratioticos* he proposed certain questions relating to ordnance, partially resolved in the second edition of this book and in *Pantometria* (1591), which entitle him to be considered as the founder of ballistical studies in England. There can be no doubt that both Leonard and Thomas Digges were extremely able men, and in their scientific attainments far ahead of most Englishmen; if Thomas was the more prolific writer, he owed what success he had achieved in the theory of artillery to 'such Scales and Theorike as my Father by his long painefull chargeable experiences in great Ordinance joyned with his Mathematicall Science first invented', the first principles he thus received being more valuable than the right to carry out experiments free of cost.[2] Digges' definition of the trajectory was like Tartaglia's, tripartite:

> The first parte of the violent course, of Gunners commonly termed the peeces pointe blanke reache, I call the Direct line of the Bullet's circuite.
>
> The second parte being a Curve Circuite, beginning at the foresaide declination from the Axis, ascending to the highest altitude over the Horizon, and ending at a like Altitude to his beginning I terme for Distinction sake his middell Helicall or Conicall Arke.
>
> The rest even to the Horizontal plaine againe I call his Declining line.[3]

The violent motion is only mechanically in a straight line, geometrically it certainly follows a curve.[4]

[1] Peter Whitehorne, *Certain Waies for the Orderyng Souldiers in Battelray and setting of Battailes* (London, 1562); Cyprian Lucar, *Three Bookes of Colloquies* (1588); Walter Ryff, *Der Furnembsten, notwendigsten der gantzen Architectur angehorigen Mathematischen und Mechanischen Kunst Weiteren inhalt des II und III Buchs der Geometrischen Buxenmeisterey und Geometrischen Messung* (Nürnberg, 1547).

[2] *Pantometria* (1591), p. 175.

[3] *Ibid.* p. 179. Cf. *Stratioticos* (1590), p. 356. 'Whether the upper part of the Circuite made with the Bullet be a portion of a circle as Tartalea supposeth. (No.). Whether it be not rather a Conical Section and different at every several Randon. (Nearer but not perfite conical, but rather helicall.) Whether it be not at the utmost Randon a Section Parabolicall in all kind of Peeces and to differ in greatnesse according to the greatnesse of the Cone that to every severall Cylinder or Peece of Ordinance is convenient, being Proportionally charged according to the perfection heretofore mentioned.'

[4] *Pantometria* (1591), p. 181.

Digges to some extent grasped the unity of celestial and terrestial mechanics—he was a decided friend to the Copernican hypothesis—and supposed that the motion of a bullet in the air is comparable to that of the planets in their epicycles, so that from a knowledge of all the different 'varietie of these Bullet's Helicall Circuites in the Ayre' taking into account the different angles of projection, and of the planes over which shots are made, it should be possible to deduce a theory for the motion of projectiles, like that of the planets, based on a system of concentric and eccentric circles.[1] Although Digges had grasped at a fundamental idea of science he was unable to make anything of it, and experience was to show that astronomy derived more benefit from mechanics than vice versa. He found a more precise analogy than the antiquated system of celestial orbs in the Archimedean spiral, formed by the uniform motion of a point along a straight line rotating uniformly about one extremity.

> So is this Artillery Helicall line of the Bullet's Circuite created onely by two right lined motions becoming more or less Curve according to the difference of their Angles occasioned by the severall Angles of Randon. Whereupon by demonstration Geometricall a Theorike may be framed that shall deliver a true and perfect description of those Helicall lines at all Angles made betweene the Horizon and the Peeces lines Diagonall.[2]

The passages in which Digges discusses the geometrical character of the trajectory offer, perhaps, the clearest prevision of the theory of projectiles created by Galileo, a theory founded on just such 'demonstration Geometricall' as Digges had hoped for, though more simply than he could have anticipated.

If Digges' object was to reveal the scientific subtleties of the art of gunnery, that of his contemporary and possible acquaintance William Bourne was to improve the practice of English gunners. Their greatest fault, oddly enough, he took to be an excessive reliance on theoretical principles manifested in their use of the

[1] *Pantometria* (1591), p. 180.

[2] *Ibid.* mispaged 168 *bis*, sig. Bb. If the size of the trajectory were comparable to that of the earth and descent under gravity uniform it would indeed be a spiral. Thus Digges' idea would seem to be, like Galileo's, an explanation of the rotation of the heavenly bodies without a centrifugal force.

quadrant without regard to circumstance. For though a gun might have such and such a range over a perfectly level and horizontal plane, this would prove of little value as a guide to practice because no stretch of ground ever satisfies these conditions.[1] No instrument could compensate for ignorance in the user of it. Bourne also complained that distances were not measured accurately. His picture of the trajectory of a projectile was curiously good, for he imagined that on leaving the muzzle of the gun the bullet flew in a practically straight line for a certain distance, beginning to curve gradually as the highest point was reached, and so more steeply down to earth again, but only if the angle of elevation was greater than 45° did the ball descend perpendicularly. He thought—and in this as in other respects the influence of Tartaglia is revealed—that the range of mortars was directly proportional to the angle of elevation.[2]

It was in the work of continental writers that Tartaglian theories received their fullest development. An early Spanish authority on artillery, Diego de Alaba y Viamont,[3] first followed the implication of Tartaglia's words by considering the hypothesis that the increase of range for each point or degree of elevation is the same, then he went on to object against a large part of the theory, especially denying that the curved section of the trajectory is an arc of circle. Rather, he said, the bullet commences to move obliquely in a very gradual way, then curves more and more steeply down until it falls vertically. From this theory of his own he concluded that the ranges of artillery are proportional to the sines of the angles of elevation; for example, if the point-blank range is 200 paces and the maximum range 2000 paces, the range at 10° will be

$$200+\left(\frac{2000-200}{\sin 45^\circ}\right)\sin 10^\circ, \text{ that is,}$$

642 paces. For mortars Alaba recommended a method of marking the quadrant so that the range increases by an equal amount for each of the ten points. According to his system if the extreme

[1] William Bourne, *op. cit.* sig. Aiii.

[2] Bourne, *op. cit.* sig. Aiii, pp. 38–41.

[3] *El Perfeto Capitan instruido en la disciplina Militar y nueva ciencia de la Artilleria* (Madrid, 1590), pp. 230 *et seq.*

range at an elevation of 45° is R paces, the range at any other angle α is

$$r = R\left(\frac{\cos\alpha}{\cos 45^\circ}\right)$$

the change from sines to cosines being necessary because mortars are fired at angles greater than 45°.

At this time, although the Germans were esteemed as the supreme exponents of the casting of ordnance, the Mediterranean countries had advanced furthest in flights of theory. Another Spanish writer of much more influence than Alaba y Viamont was Luys Collado, an engineer under the Spanish crown in Italy, whose *Prattica Manuale dell'Artiglieria* was published in both Italian and Spanish. Collado's was the first really detailed, well-illustrated technical manual on both the theory and practice of artillery, in which were set out the history and evolution of cannon as well as their employment for different purposes, and to which were added notes on such useful military subjects as mining, secret writing, and engineering. Collado, affecting to despise Tartaglia as a theorist, was not above borrowing his ideas on the passage of a bullet through the air, with the exception of the mathematical proof that the whole trajectory was curved and that the ranges of guns were proportional among themselves. Having made an experiment with a falconet, or 3-pounder cannon, he found that the ranges at the first six points of the quadrant were 268, 594, 794, 954, 1010, 1040 and 1053 paces. Consequently, noticing that the increases of range from point to point grew less towards the extreme range, he thought that it was impossible to calculate all the ranges from a single one, as Tartaglia had claimed.[1]

Another treatise in general very similar to that of Collado was compiled by Diego Uffano, who describes himself as captain of the castle of Antwerp, most of which need not be considered as, like Collado's, it is principally devoted to the founding, management, supply and tactical use of artillery. It has some importance as the first attempt to define the trajectory of a projectile mathematically.[2] For ordinary shooting Uffano explains the method

[1] *Op. cit.* (1606), pp. 117 *et seq.*

[2] Diego Uffano, *Artillerie: c'est à dire Vraye Instruction de l'Artillerie et de toutes ses appurtenances*, French translation by Théodore de Brye, (Frankfort, 1614). The original *Tratado de Artilleria* (Brussels, 1613) I have not seen. There was also a German version.

of firing by the line-of-the-metal or with a dispart, but in addition he describes the use of the quadrant in connection with his very straightforward method of working out ranges, which is to find by trial the horizontal or point-blank range R and assign the ranges for each degree of elevation as $\frac{244}{200}R$, $\frac{287}{200}R$, $\frac{329}{200}R$, etc. so that the increases in range are in arithmetic proportion decreasing regularly towards the maximum.[1]

The writings of continental authorities on gunnery from Tartaglia to Uffano were compiled and made available to English readers by Robert Norton, a gunner of the Tower, in *The Gunner*, published in 1628.[2] Norton was knowledgeable but not very consistent or accurate and his book should not be accepted as an account of English guns and gunnery at this period. His own purpose is outlined in the dedication to Charles I:

that your Majesty may be better provided hereafter of understanding gunners to manage your Artillery, the powerful regent of modern war, I have endeavoured in this practice of Artillery to supply their wants the best I can, not doubting but in short time I may work good effect therein.

Norton printed Uffano's theoretical range tables and achieved the remarkable feat of following Tartaglia in describing the curved element of the trajectory as an arc of circle and Digges in describing it as helical.[3]

It has been said of these writers of the late sixteenth and early seventeenth century that their works are of little interest, the implication being that they are now with justice utterly forgotten.[4] Examined in the light of the knowledge to be given to the world within a very few years of their publication, contrasted with the great scientific achievements of the century, indeed these early

[1] Uffano, *op. cit.* 133–5. Although Uffano was the first to print, it is doubtful if he was the first to conceive of a table of ranges. Prado y Tovar in the MS. already cited (p. 10) gives a very elaborate table of ranges at each of the six points of the quadrant for nineteen types of cannon, and tables had been transcribed by Alaba y Viamont from his sine-law of elevations.

[2] *The Gunner, shewing the whole practise of Artillerie: with all the Appurtanences thereunto belonging*. For Norton's reliance on foreign authors, whom he adapted for English conditions, see sig. B 2.

[3] *Ibid.* pp. 12, 100–1.

[4] P. Charbonnier, *Essais sur l'Histoire de la Balistique* (Paris, 1928), pp. 37–8.

works on artillery with their thoroughly medieval scientific background seem absurd and contemptible. How can their trifling arithmetical artifices compare with the imposing theorems of Galileo or the calculus of Newton and Leibniz? They reveal the fundamental uninventiveness of the mind, for during two generations at the height of the renaissance and of the rapid expansion of the use of artillery scarcely a useful footnote was added to the ballistical writings of Tartaglia, and make plain the barrenness in the field of technique, equal to the philosophical inadequacy of pre-Galilean mechanics, of the early manuals of gunnery. Although the art of shooting at long ranges was confessed to be the flower of gunnery, it was quite divorced from the physical speculations of the schools and there was no bridge between them until Galileo founded a new ballistics which derived nothing from the Tartaglian doctrines expounded by the professional experts. Yet ballistics as a science could only begin when the theory of projectiles emerged on to the field of battle from the studies of the philosophic commentators, to introduce a new and permanent characteristic to society.

It is remarkable how long the structure of the older thought lasted, how it reached from the highest to the lowest level of literary effort so that men could only see movement through the eyes of Aristotle or of his commentators. A French writer, Rivault de Fleurance,[1] occupies 67 of 192 pages with an exposition of peripatetic physics modified by the impetus theory, in the course of which he painstakingly demonstrates that all circular motion must be violent, that a body is at rest only in its natural place, that a vacuum is impossible, that 'il faut toucher pour chasser ou pousser' and other dusty treasures of the intellect. To explain continued motion he makes impetus analogous to residual heat.[2] Norton commences his treatise with a rehearsal of antiquated phrases, borrowed from Aristotle by some devious route:

> Every motion in the world endeth in repose.
>
> Every simple body is either rare and light or else thick and heavy and according to these differences it is naturally carried towards some part.

[1] David Rivault de Fleurance, *Les Elemens de l'Artillerie, concernans tant la premiere invention & theorie que la practique du Canon* (Paris, 1605).

[2] Theorem I, p. 32; II, p. 33; V, p. 39; XIII, pp. 60, 64.

Nothing worketh naturally in that which is wholly like or wholly dislike, but in that which is contrary to it and more feeble.

There is not by nature any such thing as vacuity for the avoiding of which nature maketh heavy things mount and light things descend, whereby marvellous things are performed. As we may see by our pumps which make water ascend as high as the clouds and by the Spirituals all the air is retained beneath.[1]

As late as 1639 John Roberts instructs the student of gunnery in very similar 'Principles of Philosophy fit to be known'[2] and the bankruptcy of 'official' science, the absurdities into which those who still clung to it might fall, were still apparent seven years after the foundation of the Royal Society in the compilation of Sturmy:

Every simple body is either Bright and Light or else Gross and Dark, and Ponderous, and according to the variety and difference it is always naturally carryed towards some one or other part; the World hath height as upwards or depth as downwards; and the depth dependeth upon the influence of the height.

Accident hath its variety from the subject, and goeth not from one thing unto another.[3]

These half intelligent, half decorative formulae of a system of thought that had already fallen before the attacks of Copernicus, Galileo and their successors were mingled with the spirit of trial and observation only just becoming important in pure science. Tartaglia had attempted to reconcile crude observation and philosophical principle: many times he and his followers stressed the need for measurement and the magic of numbers. Alaba's or Uffano's efforts to unravel the mystery of ranging shots resulted in wild guesses, but it was already symptomatic of the scientific mind when men began to look for the mathematical principles fundamental to superficially irrational phenomena.

It is only by giving due weight to the survival of a great (though now defeated) tradition of European thought in the Italo-Spanish school of ballisticians, by recalling that the expressions of a compiler like Norton and those who borrowed from him can be traced in an unbroken line of descent to the renaissance (Digges,

[1] *Op. cit.* pp. 3-4. [2] *The Complete Cannoniere* (London, 1639).

[3] *Mariner's Magazine*, vol. v, pp. 46-7.

Tartaglia, Leonardo), to medieval philosophy (Oresme, Buridan, Nemorarius), and so to Aristotle and the ancient world, that it is possible to explain the tardy impact of the scientific revolution. Even the most literate of gunners was neither philosopher nor scientist and the works of Galileo (1638) and Torricelli (1644)[1] passed long unnoticed in the very field where it might have been expected that their effect would be greatest and most rapid.[2] A large work of vulgarisation at the hands of the scientific academies and hack-writers had to be performed. In France Georges Fournier (1643, 1667) and in England William Eldred, Nathaniel Nye and others who obscurely profited by the wartime boom in military textbooks used the work of Uffano as a final authority, though the Frenchman had the good sense to remark that he doubted 'whether the persons who advance this rule have ever put it in practice'. Sturmy, enough of a mathematician to teach astronomy, navigation, surveying and dialling in his encyclopaedia of practical mathematics for seamen, had not yet heard of the parabolic theory. Perhaps even more surprising was the translation and publication in 1683 by Sir Jonas Moore of a mediocre Italian work of instruction in which appeared the following sentences:

> As to the several shootings in artillery, the ball being shot out flies through the air with a violent mixed and natural motion describing a parabolical line in whose beginning and end are lines sensibly straight and in the middle curved. In the beginning the impressed force driving forwards by the fire the natural gravity of the ball describeth a right line: in the middle that force diminishing and the natural gravity prevailing, describeth a crooked line; in the end the natural gravity overcoming the impressed force, which becomes weak or altogether faints, describes of a new a right line in which the ball tends towards the centre of the earth, as towards a place natural to all heavy bodies.[3]

Abstract the misused word 'parabolical' and this passage would have said nothing new a hundred and fifty years before.

[1] See below ch. IV. Roberval, Professor of Mathematics at the Collège de France, was still teaching the Tartaglian theory of projectiles about 1644, despite his close connections with the scientific movement in Italy. See his *Traité de Mécanique*, quoted by Duhem, *Études sur Léonard de Vinci* (Paris, 1906–13), vol. I, pp. 142–5.

[2] Eldred, *Gunner's Glasse* (London, 1646); Nye, *Art of Gunnery* (London, 1647, 1648); Fournier, *op. cit.* (1643), p. 130.

[3] *A General Treatise of Artillery and Great Ordnance. Writ in Italian by Tomaso Moreti of Brescia . . . translated into English by Sir Jonas Moore Kt.* (London, 1683), p. 86.

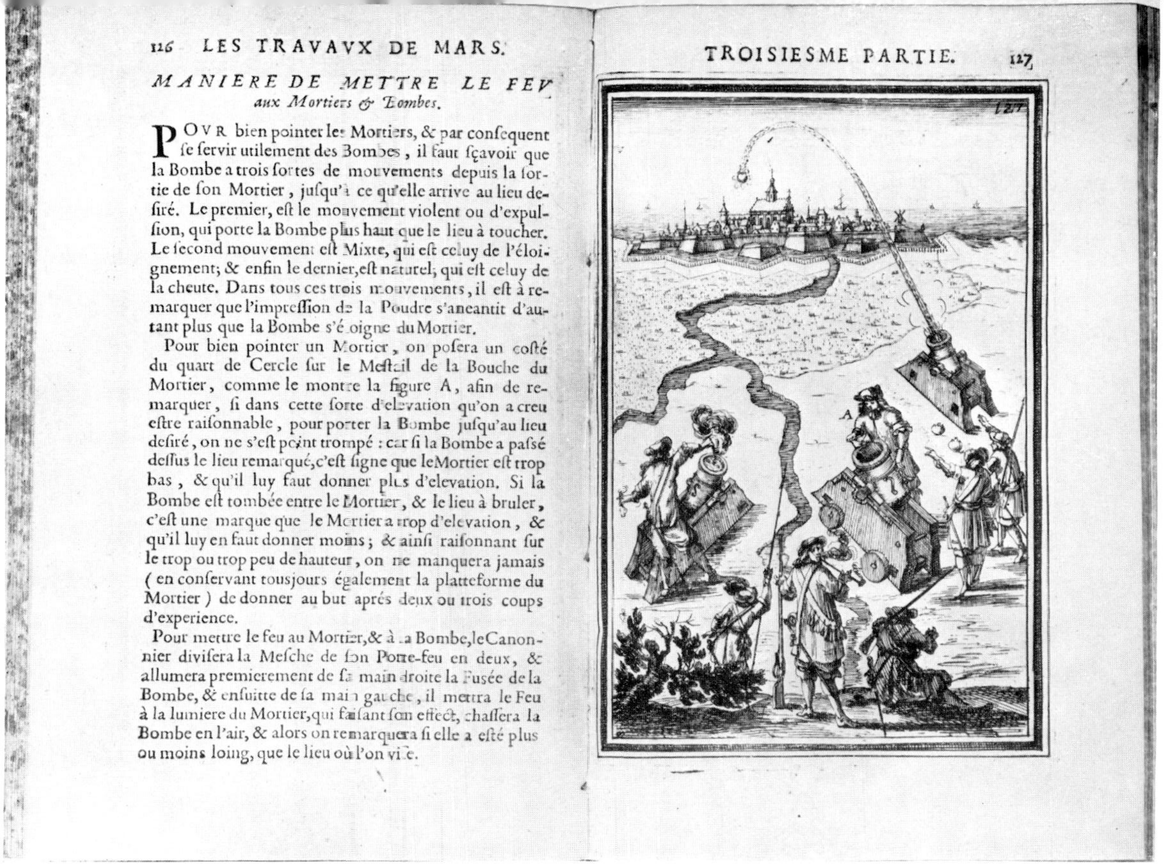

MANIERE DE METTRE LE FEV aux Mortiers & Bombes.

POVR bien pointer les Mortiers, & par consequent se servir utilement des Bombes, il faut sçavoir que la Bombe a trois sortes de mouvements depuis la sortie de son Mortier, jusqu'à ce qu'elle arrive au lieu desiré. Le premier, est le mouvement violent ou d'expulsion, qui porte la Bombe plus haut que le lieu à toucher. Le second mouvement est Mixte, qui est celuy de l'éloignement; & enfin le dernier, est naturel, qui est celuy de la cheute. Dans tous ces trois mouvements, il est à remarquer que l'impression de la Poudre s'aneantit d'autant plus que la Bombe s'éloigne du Mortier.

Pour bien pointer un Mortier, on posera un costé du quart de Cercle sur le Mestail de la Bouche du Mortier, comme le montre la figure A, afin de remarquer, si dans cette sorte d'elevation qu'on a creu estre raisonnable, pour porter la Bombe jusqu'au lieu desiré, on ne s'est point trompé : car si la Bombe a passé dessus le lieu remarqué, c'est signe que le Mortier est trop bas, & qu'il luy faut donner plus d'elevation. Si la Bombe est tombée entre le Mortier, & le lieu à bruler, c'est une marque que le Mortier a trop d'elevation, & qu'il luy en faut donner moins; & ainsi raisonnant sur le trop ou trop peu de hauteur, on ne manquera jamais (en conservant tousjours également la platteforme du Mortier) de donner au but aprés deux ou trois coups d'experience.

Pour mettre le feu au Mortier, & à la Bombe, le Canonnier divisera la Mesche de son Porte-feu en deux, & allumera premierement de sa main droite la Fusée de la Bombe, & ensuitte de sa main gauche, il mettra le Feu à la lumiere du Mortier, qui faisant son effect, chassera la Bombe en l'air, & alors on remarquera si elle a esté plus ou moins loing, que le lieu où l'on vise.

BOMBARDMENT OF A FORTIFIED TOWN BY LARGE MORTARS, 1672

The translation of Galileo's 'Discorsi e Dimostrazioni mathematiche intorno a due nuove scienze' prepared by Thomas Salusbury for his *Mathematical Collections and Translations* having been almost wholly destroyed in the Great Fire, no other appeared before Thomas Weston's in 1730. The first attempt to popularise parabolic ballistics in English was made by Thomas Venn, who published translated extracts from Galileo and also from Torricelli's 'De Motu Corporum' on the trajectory and the calculation of ranges.[1] Two years later, in 1674, the theory was very fully expounded with many elaborations by Robert Anderson in the *Genuine Use and Effects of the Gunne*, a book which became a classic for a generation. In France the parabolic theory was examined by Marin Mersenne in 1644[2] and expounded in the vernacular by François Blondel in a treatise specifically written for the improvement of artillery practice.[3] Until the third quarter of the century, therefore, the Tartaglian theory of ballistics continued almost unchallenged in the only class of books likely to fall into the hands of actual gunners. As is to be expected, the practical men were fully half a century behind the most advanced scientific knowledge.

In the same way the parabolic theory once accepted lingered on as scientific truth into the middle of the eighteenth century, despite the mathematical-physical researches of Huygens, Newton, Leibniz, Bernoulli, Robins, and Euler. The strictures of Benjamin Robins in 1742 show that the instructions given in the manuals of gunnery had not progressed beyond the ballistical theories of Galileo, the effect of air resistance upon the motion of projectiles being neglected, though it had been a favourite branch of study with natural philosophers seventy or eighty years before. Accordingly Robins, reformer and pioneer, could only declare that these writers[4] were 'very much deceived', that all their determinations concerning the flight of a projectile were 'extremely erroneous' and that the theory of gunnery was still 'in this its most important

[1] In *Militarie and Maritine Discipline* (London, 1672). He gives the Tartaglian analysis as an alternative.

[2] 'Ballistica et Acontismologia' in P. Marin Mersenne, *Cogitata Physico-Mathematica* (Paris, 1644).

[3] *L'Art de Jetter les Bombes* (Paris, 1683).

[4] Among whom may be numbered B. Forest de Belidor, *Le Bombardier François* (Paris, 1731).

branch . . . useless and fallacious'.[1] Yet these same points had already been made very much earlier by Edmond Halley in a review of a book of gunnery tables based on the parabolic theory presented by its author for the approbation of the Royal Society.[2]

An explanation of the lag of artillery training behind the science of each age may be sought in lack of education, absence of co-operation between the academists and practical men, or simply in the absence of any driving force capable of over-riding natural inertia. It is also possible to see a more specific and fundamental reason, for it may be that the ballistical theories of the text-books were unimportant and had no real influence on warlike operations, so that it was irrelevant whether or not a poor theory was replaced by a scientifically superior one. What evidence is there that the rather pretentious notions of Tartaglia or Galileo were ever used as guides to the tactical employment of artillery? In the early treatises especially, though some formal attention was paid to the mathematical parts of gunnery, the space actually devoted to them was small. There were certain pieces of information along with the philosopher's tags which the trained gunner ought to have at his command to distinguish him from a mere matross, but it may be doubted whether in a lifetime such finesse was put to the test. The real emphasis of the experts and of the artillery schools was on a sound education in straightforward field gunnery in which scientific ballistics played a very small part. This was most appropriate for the tactical use of artillery. Battles were fought at short range, both on land and at sea. Science had no place in the heat of combat.[3] It is true of course that the tactics and strategy of war were vastly modified by the introduction of firearms and that the role of the artillery in particular swelled rapidly in significance from the time of Charles VIII's descent

[1] *New Principles of Gunnery* (London, 1742); *Mathematical Tracts* (London, 1761), p. 52.

[2] *Nova Artilleria Veneta Sigismundi Albergeti Ictibus Praepollens Usu Facillime & Projectionibus Theoriae Tabularum Universalium ejusdem respondens* (Venetiis, 1703). Royal Society, *Journal Books*, 1696-1700, 26 Feb. 1700/01, 19 March 1700/01; 25 June 1701. E. F. MacPike, *Correspondence and Papers of Edmond Halley* (Oxford, 1932), pp. 167-8.

[3] The series of articles *M[ariner's] M[irror]*, vols. XXVIII, XXIX (1942-3) on 'Armada Guns' by Professor Michael Lewis provide an extremely interesting examination of the inter-relations of ordnance and naval tactics. The Spanish vessels, capable of delivering a very heavy fire at short range, were admirably designed for crushing at close quarters; this the English by keeping the wind were able to avoid, but their light longer-ranging guns were incapable of damaging the enemy (if they were so fortunate as to strike at all)

into Italy with the first of the modern artillery trains.[1] Elaborate geometrical fortifications were developed as the answer to siege artillery—again the fetish of mathematics was in evidence. But the really substantial innovations were those imposed on the planning and disposition of commanders, rather than on the training of their troops.

Warfare was still an affair of closely concentrated groups of men scarcely separated by a stone's throw, battering each other with the smallest refinement; in the last resort the decision lay in the clash of men, not weapons. Even in naval warfare a vessel was rarely entirely destroyed unless by fire on boarding. In land sieges it was usual to open the trenches and place the batteries as close as possible to the ramparts in order to obtain the maximum effect of battery, provided that this was not within 'the space that a good musket can carry point blank, or the distance at which a good marksman can hit a man, which is between 600 and 800 feet', at least until the walls were rendered untenable by the enemy.[2] Beyond this distance the besiegers would only be subject to annoyance from the artillery of the garrison, who would not waste their powder and shot on displays of random shooting.[3] From the outer trenches light cannon might be called upon to prevent the garrison becoming too active upon the walls, and mortars were capable of throwing in their bombs to crush and harass the townspeople from a distance of several hundred yards, but the serious work of cutting the breach before the storm was left to the heaviest cannon at the least possible range—'la principale batterie se fait en terrant le Canon & traversant seulement le fossé'.[4] Uffano speaks of opening 'une batterie générale' at 100-150

from a safe distance outside the range of the Spanish artillery. Only after the Armada reached Calais when the Spaniards were short of powder did the English ships close in to effective range. The lesson, very clearly, was that the short-ranged pieces and the long-ranged were equally useless beyond musket range or less—150-200 yards—and it was applied in the close broadside tactics of the seventeenth century.

[1] Sir Charles Oman, *The Art of War in the Sixteenth Century* (London, 1937), *passim*.

[2] *The Petty Papers*, ed. the Marquis of Lansdowne (London, 1927), vol. II, p. 65. The estimate of Sir William Petty agrees with others made by contemporaries.

[3] The *O.E.D.* does not suggest any connection between the two senses of the word 'random' but it is easy to imagine that shooting at random in the technical artillery sense could be confused with the sense 'scattered'.

[4] de la Fontaine, *Les Fortifications Royales ou Architecture Militaire par une Nouvelle Pratique* (Paris, 1666), p. 84.

paces.[1] The plain object of these tactics was to bring the greatest weight of metal to bear along the most direct trajectory. It is difficult to gain any idea of artillery ranges in field engagements, but they were scarcely if at all longer than those of siege warfare.

This may be confirmed by the definite knowledge, supported on all sides, that at sea nothing but point-blank fire was considered by the English commander to be worth the expense of powder, the construction and discipline of the vessel being directed towards the delivery of a heavy concentrated broadside from a distance at which no shot could fail to strike. The principal tenet of naval gunnery became: 'In a sea fight a broadside is uncertainly and for the most part ineffectually given, when it is beyond the distance of musket shot at point blank'.[2] Sir Walter Ralegh forbade any gunner under his command to shoot his ordnance at any other range than point-blank, and the same prohibition of random shooting was repeated in the Fighting Instructions of 1625, 1665, and 1691.[3] The tradition of close and brutal engagement was maintained into the nineteenth century, surviving almost until rifled ordnance removed the ineffectiveness of long-range fire. Nelson's opinion on the subject is well known:

> As to the plan of pointing a gun truer than we do at present, if the person comes I shall of course look at it and be happy, if necessary to use it, but I hope we shall be able as usual to get so close to our Enemies that our shot cannot miss their object, and that we shall again give our Northern Enemies that hail-storm of bullets which is so emphatically described in the Naval Chronicle, and which gives our dear Country the Dominion of the Seas.[4]

There is little evidence that the principles of ballistics as then understood were applied during the seventeenth or eighteenth centuries to naval warfare; on land, though the statement remains almost exact, some greater subtleties were possible. Oman records some early instances of the master-gunner marking out the ground before a battle in order that he should know how to range his

[1] *Op. cit.* p. 50.

[2] *Boteler's Dialogues* (N.R.S. 1929), written and revised 1634-43, p. 295.

[3] *Orders to be observed by the commanders of the Fleet and land companies*, etc. May 1617, Art. 26; *Fighting Instructions*, 1530-1816 (N.R.S. 1905), pp. 41, 68, 88, 126, 170, 192, etc.

[4] Sir H. Nicholas, *Despatches and Letters of Lord Nelson* (London, 1845), vol. IV, p. 292 (9 March 1801).

guns according to the dispositions of the enemy.[1] It was related of Sir George Carew at the time when he was President of Munster and engaged in the suppression of an Irish rebellion that in his attack on the castle of Rincorran, which had fallen into the hands of Spanish troops:

he performed the office of a master gunner, making some shot, and that the Artillery might play as well by night as day, himself did take and score out his ground marks, and with his Quadrant took the true level, so the want of daylight was no hindrance.[2]

Anecdotes of this type seem to be rare and the gun itself was so inconsistent in its behaviour that great accuracy in preliminary work, even in the laying of the gun itself, was labour in vain. Nothing was uniform in spite of official efforts at standardisation; powder varied in strength from barrel to barrel by as much as twenty per cent; shot differed widely in weight, diameter, density and degree of roundness. The liberal allowance for windage, permitting the ball to take an ambiguous, bouncing path along the barrel of the gun, gave no security that the line of sight would be the line of flight, even had the cannon itself been perfect. There was little chance of repeating a lucky shot since, as the gun recoiled over a bed of planks, it was impossible to return it to its previous position, while the platform upon which it was mounted subsided and disintegrated under the shock of each discharge.

Some light is thrown on the problems of shooting by the experiments of Benjamin Robins in the seventeen-thirties, for he found that at a range of some 800 yards the ball he used diverged as much as 100 yards to the right or left of the line of fire, and fell sometimes 200 yards short of the previous graze, though at a distance of 180 feet he was able to strike a small target with tolerable precision. These variations he attributed to the effects of air resistance upon the diverse motions of spherical projectiles.[3] In short there is every reason to believe with Halley that ballistical theory was of small purpose in the existing conditions of technique,

[1] *The Art of War in the Sixteenth Century*, pp. 180, 629.

[2] [T. Stafford], *Pacata Hibernia* (1633, ed. Dublin, 1810), vol. II, p. 367. For another proof that Carew had studied his Tartaglia with profit cf. *ibid.* vol. I, p. 116.

[3] *Mathematical Tracts*, vol. I, p. 150. The same points are forecfully brought out in the mid-nineteenth-century discussions of the merits of smooth-bore versus rifled artillery.

since gunners 'loose all the geometrical accuracy of their art from ye unfitness of ye bore to ye ball, and ye uncertain reverse of ye gun, which is indeed very hard to overcome.'[1]

On the other side it must be said that the increasing importance of mortar bombardment in the reign of Louis XIV did bring into wide use a technique different from that of point-blank gunnery and that this new method was far more dependent—or was thought to be—on mathematical science.[2] In the aiming of mortars there was no alternative between reliance on theory and simple trial and error. According to Blondel the latter was the French practice in his own day (*c.* 1675), the extent of the officer's skill being to increase the angle of elevation to decrease the range, and conversely. Of the demonstrative rules founded upon geometry and the theory of the motion of projectiles they were ignorant—and it is permissible to doubt whether the elaborate instruments Blondel proposed for the prediction of ranges offered the most suitable remedy for this state of affairs.[3] The difficulties of the bombardier were well summed up in a paper presented to the Académie des Sciences by a M. de Ressons in 1716, who stated that in practice he had found the theories of Galileo and his pupils to be of very little value and that there seemed to be no foundation on which calculations could be based. So many variations and accidents were possible in the loading of mortars, in the weight of the bombs, in the strength of gunpowder, 'que le plus habile & experimenté des Bombardiers ne peut répondre de tirer trois coups de suite avec justesse'.[4] Once again the lesson is emphasised that theories are vitiated by the multitude of uncontrollable technical factors.

In these two chapters I have essayed a sketch of the background of the science of ballistics at the time of its first development on the seventeenth century. It would have been possible to dwell

[1] MacPike, *op. cit.* p. 167.

[2] The great engineer Vauban was a convinced advocate of mortar fire: 'Les bombes font toujours parfaitement bien; en un mot, quinze mortiers font beaucoup plus d'effet a l'égard d'imposer que soixant pieces de canon des mieux services' (Lazard, P., *Vauban* (Paris, 1934), p. 467).

[3] Blondel, *op. cit.* pp. 3-5. Cf. de la Hire, *Mémoires de l'Académie Royale des Sciences* (1700), pp. 199 *et seq.*

[4] *Ibid.* (1716), pp. 79 *et seq.*

much longer on the field use of artillery, but this topic has been omitted as it must inevitably lead to a recapitulation of the history of warfare, which would be irrelevant to the main purpose. No one has yet attempted a general history of military techniques at this time (or much here written would have been unnecessary), and this is not the place for it. I have tried, on the other hand, to discover what guiding principles were adopted in the manufacture and employment of the most impressive weapons of war, for these are essential considerations in assessing ballistical researches. It is apparent, for instance, that governments were consumers, not producers; they took freely of what was available to them in men or material, used them for their own ends with a minimum of effect upon tradition, and restored to the community very little by way of knowledge or experience. As a discerning patron of intellectual endeavour of any sort the state was negligible, nor were the problems it might present to science clearly appreciated. It still depended upon the technical resources of private craftsmen; it had not yet discovered that techniques could be moulded to suit its own narrower ends of policy. Failing to expand the resources of the state qualitatively as well as quantitatively, governments matched their methods to their powers, placed their trust in the pressed man and the amateur in preference to training and research. The enormous requirements of administration and war, compelling them to buy what the craftsman could produce quickly and cheaply, gave the advantage to routine and mediocrity. As always, the virtues of size were more obvious than those of skill, and the directive of the state was all towards improvements in scale, not in type.

The contrast between the neat precision of theory and the clumsy confusion of practice is clear and sharp. Artillerists—first of all craftsmen to take up a cloak of science—struggled through many generations with one and another set of dogmas based upon natural philosophy, in search of the mathematical formulae to rule their art, while always their tools remained undeveloped, incapable of using the worst of their theories. But simple reason and experience were not enough for those who had faith in a fundamental '*theorike*' of mechanics.

Something, too, can be learnt from the history of guns and

gunners of the wider scene of the scientific renaissance. The really constructive minds were bright stars shining in a not altogether blank sky, for the work of popular mathematical education went on, in many obscure hands. Gunnery, like seamanship, surveying and architecture provided substance for the operation of many a calculating, mechanician's mind. In these fields a quasi-scientific ability could be cultivated, and the advantages of method dimly apprehended. For a sincere, if limited, exposition of the virtues of the rational approach one need look no further than Robert Norton:

... for the better understanding of the sequent discourses, we shall do well first to conceive that every material thing is either to be lineally described, or else intellectually understood by some proper figure, or apt word, name or definition, which properly belongeth thereunto. For as every art hath certain rules and principles (to proceed) without the knowledge of which no man can ever attain unto a necessary perfection for practice thereof, unless he first endeavour to learn (rather by reason than by rote) what each part thereof is, with the name and nature of each member and part of it. ... The neglect of which is the cause why many (otherwise well affected to Art) do so fruitlessly bestow their time, labour and cost to no purpose, often condemning the Art as too hard for them, when (God knows) the only cause is their disorderly progress in the study and practice thereof. ...[1]

The gunner, in short, could rise to be a workaday member of one of those groups which existed on the fringe of science, nourishing himself and his studies on the crumbs from philosophic feasts. Recruited from the hard service of the seas, from the impoverished gentry or a line of martial ancestors, such men as these, with the intellectual cream of other crafts, were the aptest interpreters of science and the discoverers of its utilitarian charms.

[1] *The Gunner*, sig. B v.

CHAPTER III

INTERNAL BALLISTICS

The early artillerists did not confine their ballistical studies to the motion of the projectile but tried also to understand what was happening to the gun itself at the moment of firing. Scientific attempts to investigate the burning of explosives and to measure the force of expanding gases, to inquire minutely into the metallurgy and shape of the cannon, have only been made in recent times; there were already in the sixteenth century gropings after the objects of modern research through trial-and-error modifications of the length of the piece and improvements in the quality of gun-metal to enable it to withstand heavier charges. At the same time the chemical composition of gunpowder and the methods of its manufacture were examined with a view to rendering it more powerful, uniform, and cheap.[1] Just as the new discoveries in physics and mathematics made possible the foundation of exterior ballistics by Newton, Leibniz, and Bernoulli, so the progress of chemistry provided the basis for an understanding of the phenomena of combustion and explosion, although, because chemistry was far behind the physical sciences, its dependent branches were primitive in comparison with mechanics, and no significant changes took place in the chemistry or metallurgy of artillery before the nineteenth century.[2] The theorems of motion propounded by Galileo themselves helped to clear the path for the study of internal ballistics by defining the scope of each branch of the science, for only when velocity and angle of projection were recognised as the two single factors affecting the range of a gun was it feasible to distinguish between the changes brought about by alterations in the one or the other.

[1] The defeat and destruction of the French fleet at La Hogue in 1692 was attributed by its commander partly to the weakness and insufficiency of the powder supplied, leaving him to be out-ranged by the English (La Roncière, *Histoire*, vol. VI, pp. 100, 125).

[2] Gunpowder was strengthened in the late eighteenth century by the more effective carbonisation of wood in iron retorts (Douglas, *Naval Gunnery* [1820], p. 201).

From the first days of artillery founders had striven to make guns safer to friends as well as more fatal to enemies. It was necessary to discover why, from time to time, they burst without any apparent negligence on the part of the gunner, as it was also to reduce the vibration and recoil which shook to pieces their carriages and disintegrated platforms and ramparts. Further, the natural guess that the violent bounding of the piece might be a major cause of errors in shooting suggested that the recoil of artillery ought to be studied.

Just before the beginning of the century powder had been much increased in strength by innovations in manufacture. The most important of these was the discovery that though the ingredients of the powder must be very finely ground and closely incorporated, the explosive power was greater if it was not left in this fine state (called serpentine) but 'corned' into grains of various sizes by rubbing it while damp through a sieve. Corned powder, more quickly burning and stronger, had to be used with discretion and was responsible for the tendency to cast artillery of greater weight in proportion to its calibre or, as they said, with more fortification, and to reduce charges, between about 1580 and 1620. Like all complications, the new process favoured standardisation. In both France and England the explosives industry was jealously regarded by the state and carefully controlled. Free manufacture or sale were not permitted and the consignments from contractors were regularly checked.[1] It also became advisable to specify the quantity of powder to be used in each type of gun, despite the mitigating effects of the allowance for windage (which wasted from a quarter to a half of the powder put into a piece); as the improvements in gunpowder outstripped those in gun-founding, this was reduced from a weight equal to that of the shot to between two-thirds and a half in the seventeenth century, and a third or even less in the eighteenth. The limiting factor in the attainment of high velocities of projection was not so much the poor quality of the explosive as the wasteful design and feeble structure of the gun. Even had higher muzzle velocities made longer ranges possible, the gun was too inaccurate to make effective use of them. As

[1] C. G. Cruikshank, *Elizabeth's Army* (London, 1946), p. 62; *Ordnance Minutes* W.O. 47 *passim*.

usual, advances were more necessary in the practical than in the scientific aspects of gunnery.

Saltpetre was early appreciated as the vital component in the mixture of substances making gunpowder. Possibly this was a conclusion drawn from the spectacle of the lively sparkling of saltpetre when flung on to glowing coals; at any rate for some such empirical reason sulphur and charcoal were looked upon as necessary but auxiliary chemicals. The chemical knowledge of Tartaglia was able to go only so far into the nature of explosives as to explain that each substance supplied a defect in the others,[1] but he knew that the powder was stronger in proportion to the amount of saltpetre.

Whether the various recipes that Tartaglia collected are historically exact or not, he was right in believing that experience has taught the value of increasing the proportion of saltpetre in gunpowder.[2] Other ways of augmenting the violence of explosion were tried, especially the addition to the ordinary powder of chemicals prominent in the pharmacopœia such as antimony, mercury, vinegar, and alcohol. Nor was gunpowder the only explosive known after the discovery of fulminating gold about the end of the sixteenth century, followed by Künckel's discovery of fulminate of mercury about a century later.[3]

In England, during the first years of the Royal Society's existence, its chemists devoted some time to the consideration of explosives. Prince Rupert was reported to be the inventor of a gunpowder ten times as strong as that ordinarily made; its ingredients were those always used but prepared with more particular attention to their purity, fineness, and incorporation.[4] In a series of experiments a comparison of the strength of *aurum fulminans* and ordinary gunpowder was attempted. One product of these trials was a new type of machine for testing powder made by Robert Hooke. Such instruments had been used since at least the mid-sixteenth century with the common principle of a small chamber fitted with a touch-hole and a strong lid, in

[1] 'Cadauno de loro media e suplise ad alcun diffetto de alcun delli altri dui.'

[2] *Quesiti et Inventioni* (1551), pp. 40, 42–3.

[3] *Collegium Physico-chymicum Experimentale oder Laboratorium Chymicum* (Hamburg und Leipzig, 1716), pp. 213 *et seq.* Cf. note by Tenney L. Davies in *Isis* vol. x, p. 167.

[4] Birch, *History of the Royal Society*, vol. I, pp. 281–5, 292.

which a few grains of powder were fired, forcing up the lid against the resistance of a spring or gravity, and the goodness of the powder was computed from the height it reached. Many experiments were made by the Society on the strengths of various compositions of gunpowder with this machine of Hooke's and others.[1]

More closely appertaining to the pure theory of chemistry were the different explanations offered of the primary phenomenon of internal ballistics—the rapid expansion of the gases released from the powder which eject the shot with a loud report. Biringuccio believed that according to the teachings of natural philosophy 'the power of the elements is in the simple parts that the powder is composed of'. The root of the matter he took to be a 'certain subtle dryness ready to introduce fire easily' and multiply it in a certain proportion which philosophers have discovered 'con sperientia'. As fire takes up ten times as much room as air, air ten times as much as water, and water ten times as much as earth, when the powder, which is a corporeal, earthy thing, is turned into fire, air and the moist, subtle earthiness of smoke, it needs much more space and bursts out at the weakest point.[2] A similar 'explanation' based upon the Aristotelian system of the elements is given by Rivault, who points to tremendous natural manifestations of the power of expansion, such as earthquakes caused by the heating of the water in the earth's interior.[3]

The notion that saltpetre contains air within itself liberated by the burning appears to be a good deal older than the ideas of Boyle, Hooke and Mayow, for Robert Norton wrote in 1628 that the salt was a 'body of air transformed into earth, apt by moisture to be dissolved and by fire to be resolved into his first air state, being an airy substance fixed in dry earth'.[4] This is about as clear a statement as any other put forward from the

[1] Birch, *op. cit.* vol. I, p. 302; Robert Gunther, *Early Science in Oxford*, vol. VI, pp. 152–3. William Bourne, *Inventions or Devises very necessary for all Generalles and Captaines or Leaders of men, as well by Sea as by Land* (London, 1578), no. 54, p. 39. On 25 April 1708 there was an experiment with a different explosive; five parts of nitre and three of double-refined sugar 'by a flash from a flint and steel were both fired and burnt as gunpowder. Mr Ayres who suggested this experiment said he had tried this powder in mortars, muskets and pistols and that it had the same effects as gunpowder' (*Journal Books*, vol. X [1702–14], p. 183).

[2] *Op. cit.* bk. X, ch. II (1943 ed., pp. 410–11).

[3] *Op. cit.* pp. 126–33.

[4] *The Gunner*, p. 142.

seventeenth century as 'foreshadowing the discovery of oxygen'. Descartes also analysed the explosion of a charge of gunpowder, rightly stressing the importance of saltpetre and characterising explosion as extremely swift combustion and expansion.[1] Boyle applied the corpuscular doctrine, regarding explosion as the rapid disintegration and agitation of the particles of matter in the gunpowder. When fire is applied the grains

> are suddenly fired; that is, many millions of parts which before lay still and at rest, are by the action of the burning coals shattered as it were and put into a posture ready to be agitated by the rapid motion of the *materia subtilis*: into which posture they are no sooner put, than agitated and whirled sufficiently by it: whence follows a vast expansion of that part of gunpowder so fired.[2]

Both Hooke and Robert Boyle were among the members of the Royal Society interested in experiments to measure the volume of 'air' produced from gunpowder by burning it in vacuo.[3] It would be enlarging on this topic too much to discuss all their researches into combustion, or into the chemical composition of nitre in which the burning of gunpowder played an important part, but it was sufficiently proved that saltpetre contained whatever was lacking to enable a combustible to take fire.[4] Similar researches at the Paris Académie des Sciences were carried out by Denis Papin under the direction of Christiaan Huygens, and after his move to England in 1678 Papin continued and extended his work for the benefit of the Royal Society.[5]

In one of his papers submitted to them he wrote 'I should therefore make no difficulty to believe [sic] that all the effect of gunpowder comes from the air which is compressed therein and especially in the saltpetre, for I have not observed that brimstone yields air'.[6] It is worth noting that Papin designated the product of combustion as 'factitious air' and had discovered that 'it could

[1] *Principia Philosophiae*, pt. IV (1644), ed. Adam-Tannery, vol. VIII, pp. 263 *et seq.*

[2] *Works* (1772), vol. I, p. 181.

[3] Gunther, *op. cit.* vol. VI, p. 292.

[4] Gunther, *op. cit.* vols. VI, VII, *passim.* T. S. Patterson, 'John Mayow in contemporary setting'. *Isis*, vol. XV (1931), pp. 47 *et seq.*

[5] *Nouvelles Experiences du Vuide* (1674); Huygens, *Œuvres*, vol. XIX, pp. 236 *et seq.* The study of explosives was included by Huygens in a programme of research of 1666 (*ibid.* p. 256).

[6] *Philosophical Transactions*, no. 122, 21 Feb. 1675/6, p. 548.

be of no use towards producing sparks'.[1] He had, of course, no knowledge of the function of oxygen in combustion nor indeed any clear idea of a gas so that he was quite unable to distinguish between the various gases yielded by the burning gunpowder and the oxygen or 'air' which he imagined to be released from the saltpetre, leading him to the conclusion that all the 'air' produced was at first present in the potassium nitrate, the sulphur and charcoal contributing nothing. It is particularly interesting that the violence of explosion caused by expanding gas should have received such a typically mechanical, rather than chemical, explanation. The firing of gunpowder was treated by Papin and Hooke as analogous and strictly comparable to the sudden release of compressed air in a wind-gun—a toy which was now sufficiently familiar.[2] Whereas the air in the wind-gun was compressed to one-tenth or one-twentieth of its normal volume, it was in a vastly more compact state in the particles of saltpetre, from which it was released by fire, but mechanically the results were the same. That energy might be derived from a change of chemical state was a concept yet unformed. Yet Papin's ideas represent the culmination of the chemistry of explosives in the seventeenth century.

Another point in the phenomena of artillery which suggested itself for scientific consideration was the measurement of the force of gunpowder, a problem which could be approached in various ways. One method would be to investigate the effectiveness of a gun by ascertaining its ranges under fixed conditions or the penetration of its shot into different materials, comparing these these results with those obtained by mechanical contrivances. This would give a sort of mechanical equivalent for the work done by a certain weight of powder acting upon a certain weight of projectile. But after the formulation of the elementary principles

[1] Experiments of 26 Jan. 1686/7, 13 April 1687. Royal Society, *Classified Papers*, 1660–1740, vol. XVIII, Papin, nos. 41, 50.

[2] It is worth while to recall that Papin continued in England his efforts to make a workable gunpowder engine of the type proposed by Huygens who used the explosive to produce a partial vacuum in a cylinder, and the weight of the atmosphere. It was in introducing steam instead of powder than the atmospheric engine was invented. (Cf. Huygens, *Œuvres*, vol. VII, p. 356 (1673); vol. VIII, p. 482 (1684); vol. IX, p. 235 (1686), p. 465 (1690); *Acta Eruditorum*, Aug. 1690; *Divers Ouvrages de Mathématiques et de Physique par Messieurs de l'Académie Royale des Sciences* (Paris, 1693), pp. 320 *et seq.*; Royal Society, *Classified Papers*, *loc. cit.* no. 65 (read 26 Oct. 1687); *Register Book* (copy), vol. VII, pp. 24 *et seq.*, etc.

of mechanics by Galileo it was realised that measurable performance depended solely upon the velocity and mass of the projectile, and that therefore the primary necessity for the rational study of ballistics was a measurement of initial velocity. The first even approximate measures were only obtained by Benjamin Robins in the early eighteenth century with his ballistic pendulum; earlier attempts to see how quickly a bullet passed over a distance of some hundreds of feet were no more than guesses.[1]

Some 'Experiments for trying the force of Great Guns by the learned Mr Greaves' (probably John Greaves, 1602-1652, Savilian Professor of Astronomy at Oxford) were published in the *Philosophical Transations* for 1685, and may be an indication of a revival of interest in military studies at that time since the paper was written thiry-four years earlier. Greaves had set up three butts of massive oak at Woolwich and fired at them different projectiles with various charges. His best shot was from a brass culverin of 5300 lbs weight which shot an 18 lb. ball quite through the first two butts and lodged in the third.[2] These experiments do not seem to have been repeated, but the Royal Society took up the question of the velocity of projectiles from firearms. Robert Hooke reported to Boyle that the apparatus designed for the experiment had failed to function and no results were obtained, although some odd effects were noted. A carbine had been loaded and a thread fixed across its mouth which, when broken by the bullet, would release a pendulum. After a certain distance the bullet was to strike a board and in so doing by an arrangement with taut string stop the swinging of the pendulum, whose vibrations would measure the time of the bullet's flight over the determined distance. Somehow the bullet failed to break the thread and passed right through the board at the far end.[3]

Order for the repetition of the experiment was given, and in September 1664 Hooke and Dr Charleton reported that 'as near as they could observe' the velocity of a bullet shot from a musket with Prince Rupert's powder was 'above six score yards in half a

[1] Described in *New Principles of Gunnery*. Robins was able to use only a small machine and a bullet of an ounce or two in weight. The first trials with large projectiles were made by Charles Hutton (see his *Mathematical Dictionary*, London, 1796).

[2] *Phil. Trans.* no. 173, July 1685, pp. 1090 *et seq.*

[3] 25 Aug. 1664. Gunther, *op. cit.* vol. VI, pp. 188, 191.

second'—rather a low estimate, perhaps. The Society was still not satisfied with the accuracy of the result, but like so many endeavours, this task was laid aside, for Hooke, indefatigable experimenter though he was, had too many other calls on his ingenuity. Barometric experiments, optical researches, measurements of the speed of falling bodies, dissection of a viper, miscroscopic observations, all were receiving his attention at this time and from among his work he had to select one experiment a week for the entertainment of the Society.[1]

The first useful experimental study of recoil, made by Lord Brouncker at their request in 1661, must also be attributed to the lively scientific and technical curiosity of the Royal Society. The point at issue was whether or not the recoil began before the projectile was shot from the piece, and further, if it did, whether the recoil could affect the direction of the shot's flight. Brouncker was able to answer both questions affirmatively, though he found the problem less simple than had been expected.

The public interest of these ballistic researches may be judged from the places where they were made—firstly in the court of Gresham College in the presence of the Society, and later in the Tiltyard at Whitehall before Charles II and his brother.[2] Brouncker's apparatus was a musket barrel attached to a triangular frame with its breach in the centre of the base. The frame was allowed to recoil when the charge in the barrel was fired by pivoting round one or other of the base angles (which was firmly held), thus swinging the muzzle round in an arc. He found that with small charges of powder the bullet always shot to that side of the mark (towards which it was pointed) to which the barrel swung during recoil, left or right as the right or left corner of the base was fixed; but if the charge was increased to more than some forty-eight grains in weight the bullet flew to the opposite side of the mark, that is, struck on the side of the fixed corner. The first effect was readily explained by supposing the ball to be carried

[1] Gunther, *op. cit.* vol. VI, pp. 192, 202.

[2] The experiment was first performed on 7 April 1661 and the account of it presented on 10 July. Brouncker had been reminded of his promised demonstration three times since Jan. 1660/1 (Birch, *op. cit.* vol. I, pp. 8 *et seq.*). The account is printed in Thomas Sprat's *History of the Royal Society* (London, 1667), p. 237; and from another copy in the correspondence of Huygens and Sir Robert Moray in Huygens, *Œuvres*, vol. III, p. 323, Cf. also vol. III, p. 287; vol. IV, pp. 35 *et seq.*

round by the barrel in its recoil-swing; to account for the contrary effect with larger charges Brouncker assumed a more complicated motion. Because the mouth of the barrel is moving sideways at the instant of discharge, the wall of the barrel must be contiguous to the bullet, therefore the last blast of the powder will be altogether on its inner side, sending it off in the opposite direction to that of the recoiling barrel:

wherefore the bullet must necessarily cross the axis of the piece and that with a greater or lesser angle according to the force of the powder. And when this angle, therefore, is greater than the angle of Recoil then must the Axis of that Cylinder in which the bullet moves cross the axis of the mark, beyond which intersection the mark being placed the bullet must be carried necessarily wide of the mark of the contrary side to the recoil of the piece.

From a practical point of view Brouncker's investigations are very interesting, for they showed how wide and unpredictable might be the shooting errors of a smooth-bore gun even at low velocities and short distances, especially if recoiling over rough or sloping ground, but the mathematical formula he arrived at to embrace his results remains incomprehensible.

Brouncker also devised an experiment to demonstrate the principle which Newton was to formulate as his Third Law of motion, namely, that action and reaction are equal, which was much improved upon by Edmé Mariotte. Having propounded the rule that if two elastic bodies impact directly with velocities reciprocal to their weights each body will return with the same velocity, Mariotte added as a corollary that this reasoning could be used to explain the recoil of cannon and other firearms. For if a small mortar is loaded with a ball equal to one-eighth part of its weight, and placed horizontally, when fired the expansion of the flame ('le ressort de la flamme') will act on both mortar and shot as a spring on unequal weights and they will fly apart with velocities reciprocally as their weights, i.e. the velocity of the ball will be eight times that of the mortar.[1] This proposition he illustrated very neatly. A cylinder of lead was made to fit loosely into a pistol barrel and both were hung from parallel threads close together and ten feet long, suspended from nails in

[1] *De la Percussion ou Choc des Corps*, Ie partie; *Œuvres* (Leyden, 1717), vol. I, pp. 29, 33.

a wall. The pistol was weighted with lead until it was just five times as heavy as the cylinder which was inserted into it as a projectile. When the charge in the barrel was fired the two flew apart, the angle of recoil of the barrel and of elevation of the cylinder being measured on the wall, which was marked in degrees. The results obtained, when allowance had been made for the effect of air resistance, which reduced the velocity of the cylinder progressively more than that of the pistol as they were shot further apart, showed that Mariotte's proposition was true within the limitations of experiment.[1]

Tabulated results of Mariotte's Experiments[2]

No.	*Angle of Recoil*	*Angle of Elevation*	*Ratio* (V_1/V_2)
1	$9\frac{1}{2}°$	$47°$	4·817
2	$5\frac{1}{4}$	26	4·968
3	$16\frac{1}{4}$	82	4·644
4	$13\frac{1}{8}$	67	4·697
5	8	$40\frac{1}{4}$	4·940

Some further points relating to internal ballistics found in the older 'practical' writers and originating like so much of their doctrine with Tartaglia are worth mentioning. It was often asked why the first shot from a cannon did not range so far as the others. Tartaglia saw the reason for this in the disturbance of the air set up by the first round, creating as it were a favourable eddy to hurry the next shot along, while the heating of the gun helped to fire the powder more quickly.[3]

Gun-founders were particularly interested in correctly proportioning the length of the bore to the calibre. It was impossible (and almost certainly it would have been entirely useless) to cast the very heavy cannon relatively as long as the lighter pieces, so that whereas the cannon class were up to twenty calibres long, and periers even shorter, the culverin class were made with a

[1] *Œuvres*, vol. I, p. 35.

[2] I have calculated this column on the assumption that the relative velocities V_1, V_2 of the cylinder and the pistol are $\frac{V_1}{V_2} = \sqrt{\left(\frac{1-\cos\theta_2}{1-\cos\theta_1}\right)}$.

[3] *Quesiti et Inventioni*, bk. IV, p. 13.

bore thirty or more calibres in length. Exact dimensions could only be discovered with experience, but it was generally allowed that range increased in proportion to the length of the barrel.[1] Shorter guns had the practical advantages of lightness, cheapness and convenience of handling, especially on board ship, hence the practice of 'cutting' the culverin. Tartaglia declared that there was a certain optimum length of barrel for a particular charge and projectile, and that if any cannon was either much longer or much shorter than this it would not shoot so well. This opinion was well in accord with philosophic notions of fitness. The correct length of a gun he defined as being such that 'in that instant when all the powder has been resolved into fire, in the very same the ball shall have arrived precisely at the end of the bore, that is at the mouth of the piece', for in this way the shot is subjected to the greatest violence of the powder, none of which is wasted unburnt, and a longer barrel would only impede the flight of the projectile.[2] Cardano tried to solve the same problem in his book on proportions;[3] other writers sought to work out scientifically the optimum charge for a given projectile.[4]

The most characteristic feature in the design of early artillery was the large allowance to be made for the windage of the shot, a technical subject discussed very fully by the standard authors such as Collado and Uffano. As a safety-valve the windage had a double purpose, lessening the danger that a large, rusty or irregularly cast shot might stick in the bore, and providing an outlet for the powder gases to reduce the pressure on the gun, whose strength, even after proof, remained very uncertain. Various methods of calculating the difference in diameter of bore and shot were followed; a geometrical method giving a windage of one-twenty-second part of the diameter of the bore was popular, or the fraction 1/21 was suggested to those more arithmetically proficient. One-twentieth of a calibre was the standard windage of English artillery in the Napoleonic period, which gives a windage of a quarter of an inch for a five-inch diameter ball,

[1] Cf. *Pirotechnia* (ed. 1943), p. 224.

[2] *Quesiti*, vol. II, pp. 19–20.

[3] *Hieronymi Cardani Mediolanensis . . . Opus novum de proportionibus numerorum motuum ponderum* (etc.), lib. V, Prop. CXVI (ed. Basileae, 1570), pp. 111–12.

[4] *Quesiti* vol. IX, pp. 19, 21. Rivault, *op. cit.* p. 70.

and in reality when an undersized or flattened shot was used the gap might be much larger.[1]

This very great obstacle in the way of accurate shooting based upon scientific ballistics naturally remained until Armstrong and Whitworth's rifled ordnance drove out the smooth-bore gun, although already in the seventeenth century there were a few advocates of the almost unrealisable project of using nicely fitting, machined guns and shot among the few scientists who had engineering sense. Halley informed the Royal Society (2 July 1690)

> that the fitness of the shot to the bore of the piece was of great consequence in gunnery . . . that by observing this more powder might be saved than would pay for the turning of our great cannon shot, that another great advantage arising from it was that a shot could be made with much more accuracy, and a third that guns need be neither so long nor so weighty as are now in use and yet do the same execution.

He told how he had seen a shot of 14 lb. cast more than 550 yards from a barrel ten inches deep with two ounces of powder only because it was well fitting. Some months later he reported on the improvement in practice that he had obtained in experimenting with a tight-fitting tompion or cylinder of wood placed between the charge of powder and the shot.

Another inventor was the Italian Sigismondo Alberghetti, who proposed to use an ovoid projectile to overcome the irregularities of the flight of an imperfectly spherical shot, particularly its random rotation in the air caused by a bouncing passage down the barrel. All such projectiles were rendered fruitless by the lack of suitable machine tools, which were only designed about the end of the eighteenth century, though it is interesting to find Alberghetti praising Prince Rupert for his attempts in this direction, and submitting his book to the criticism of the Royal Society.[2]

With the exception of Tartaglia's discussion of whether or not there are optimum sizes and charges for cannon, which was based entirely on *a priori* notions of what would or would not happen in certain circumstances, there was little interest in internal ballistics compared with the attention given to the elucidation of

[1] Cf. Sir Howard Douglas, *op. cit.* (ed. 1817).

[2] *Nova Artilleria Veneta* (Venice, 1703), p. 4.

the mathematical-physical principles of the theory of projectiles. Nor is this unnatural, since the latter had played an important part in physical discussions since the time of Aristotle, had been discussed with approval or refutation of peripatetic doctrines by every commentator for two millennia, and had became of apparent practical importance when firearms became capable of discharging missiles at considerably greater ranges than had been attained by classical or medieval military engines. The type of problem now included in the field of internal ballistic rarely had the same general philosophic interest and was tackled by the craftsman in his own empirical way. Moreover, while mathematics at the end of the century was tolerably well equipped to analyse the trajectory of a projectile in a resisting medium, metallurgy and chemistry were still in too rudimentary a stage of development to do better than offer more or less acute guesses for the improvement of ordnance. As a result great advances were made in theory which were of little practical use and had negligible effect on the actual employment of artillery, while the mechanical improvements, far outrun (as Halley had seen) by theory, and essential to any radical change in the character of warfare, were neglected. It was the engineering ingenuity of the nineteenth century, not the progressive elaboration of dynamical theories originating with Galileo and developed by Newton, that was responsible for the revolution which then, at last, occurred, with the introduction of scientific ballistics to gunnery. Although seventeenth-century science had been able to throw a brilliant light upon a few scattered topics, it had failed to create the synthesis of craft knowledge and natural philosophy which some, like Robert Boyle, had believed to be both desirable and possible.

CHAPTER IV

MATHEMATICAL BALLISTICS I

Introduction

At the end of the seventeenth century it was possible for those who befriended the new experimental philosophy to point to a list of names of men of such genius in science as had graced no previous century, from Galileo and Gilbert at the beginning to Huygens and Newton at its close, and to recount a series of discoveries which would stand comparison with any recorded triumphs of human thought. In every branch of science the frontiers of knowledge had receded, some outposts of ignorance and superstition had yielded. The development of each showed the influence of a master, so that it had become a commonplace to speak with pride of Galileo and astronomy, Harvey and anatomy, Boyle and physics.[1]

In mathematics the progress which had been made in the period from Galileo's prime to that of Newton was no less astonishing. Leibniz might claim with justice that his differential calculus was as great an advance on the analysis of Vieta and Descartes as that had been on the methods of antiquity.[2] It is a characteristic of this period that a number of its great philosophers are also figures of considerable stature in the history of mathematics. Another reflection is perhaps less obvious: mathematics had penetrated into

[1] Sir Hugh Platte in the *Jewell House of Art and Nature* (London, 1594), asks 'Why then should we think so basely of ourselves and our times? Are the paths of the ancient philosophers so worn out or overgrown with weeds than no tract or touch thereof remaineth in our days whereby to trace or follow them? Or be their labyrinths so intricate that no Ariadne's thread will wind him out that is once entered?' (Sigs. B1–B2).

From a very different source, compare the words of John Wallis to Leibniz (20 April 1699). 'Praesenti seculo (quod jam ad finem vergit) eruditionem in omni rerum genere insignes (et quidem insperatos) processus obtinuisse, certum est; in re Physica, Medica, Chemica, Anatomia, Botanica, Mathematica, Geometrica, Analytica, Astronomia, Geographica, Nautica, Mechanica, ipsaque (quod minus laetor) Bellica, et quidem longe majores quam per multa retro secula obtinuerint' (K. I. Gerhardt, *Leibnizens mathematische Schriften*, Band IVIII [1859], p. 66).

[2] Huygens, *Œuvres Complètes*, vol. VIII, p. 215.

many activities where previously judgement, experience or taste alone had prevailed.[1] There was a strong and growing tendency to replace human fallibility by the infallibility of numbers and formulae. Four principal reasons may be suggested for this: the inquisitiveness which, giving the name of *curieux* to a whole generation of scientists and amateurs, sought to find the precise laws behind every aspect of nature; the influence of the modern state with its growing interest in statistics and accounting; the widespread belief in the existence of unsurpassable standards of excellence, subject to rigid rules, expressing itself alike in (for example) classicism in art and literature and the striving after optima of mechanical performance; and lastly the increasing depth and flexibility of mathematics, the development of the instrument itself to cope with complex questions which could only be resolved in equations of many variables. Moreover the publication of innumerable popular manuals for surveyors, merchants, seamen, soldiers and others in whose walk of life simple mathematics might be applied (the work of those rather mysterious figures 'teachers of the mathematicks' who were already flourishing in the commercial capitals of Europe) was both a symptom and cause of this process.

While the application of mathematics to human needs was nothing new where the usefulness of measurement had been obvious from time immemorial—as in land survey,[2] inheritance, and the management of commercial exchanges,[3] it was now extended to matters where the existence of relations capable of mathematical treatment was not at once apparent; and to the more enthusiastic writers the limits of the utility of mathematics were not yet in sight. The opinion shared by Galileo, Mersenne, Gassendi, Descartes, Huygens, Leibniz and Newton that natural philosophy must remain vague and even dependent on occult explanations until it was brought under mathematical discipline

[1] G. N. Clark, *The Seventeenth Century* (Oxford, 1947), p. 233.

[2] Providing an example of assessment without geometric measure, since the early unit was based on time, the day's work.

[3] 'L'histoire scientifique du moyen âge montre d'ailleurs que ce sont des questions de commerce et de finance qui ont maintenu et même rendu florissant, surtout en Italie, l'enseignement du calcul et de l'arithmétique, jusqu'au moment où la renaissance de l'astronomie introduisit de nouveaux procédés' (Paul Tannery, *Mémoires Scientifiques* vol. x, p. 27.

is too well known to require emphasis, while even Boyle, writing of the 'Usefulness of mathematics to Natural Philosophy' regretted that he had not studied the subject more seriously in his youth.[1] In the field of practical arts the use of calculation was either introduced or improved. Surveying was rendered more exact by the invention of new instruments, such as the theodolite, first described by Digges in 1571, and the levelling instruments of Huygens, Thevenot and others. The traditional craft of ship-building was examined afresh in the light of the knowledge that the shape of the hull and the disposition of the rigging, upon which the shipwright relied to obtain speed and seaworthiness in his vessels, could be subjected to mathematical analysis;[2] in architecture the rational study of the strength of materials begun by Galileo formed a new branch of mechanics, and it was currently believed that the finest artistic effect could be obtained only by strict adherence to rigid rules of proportion;[3] the theory of music was enlarged by the discovery of the simple mathematical rules connecting the lengths and tensions of strings sounding in harmony;[4] even such humble techniques as that of gauging, or measuring the volume of barrels and vats which was of some importance to the customs officers, were given a mathematical logic.

The new outlook in mathematics, becoming more widely accepted during this utilitarian century, was reflected not only in its books but in its educational system. It became quite clear that in some of the professions the man who hoped to rise by merit to an eminent position must be familiar at least with the elements of geometry and algebra, as well as simple arithmetical operations. The same skill was frequently urged on the ambitious artisan by publicists. A new type of education arose to supply the need of men like Phineas Pett, master shipwright, who spent

[1] *The Usefulness of Natural Philosophy. The Second Tome Containing the later section of the second part.*

[2] The Marquis of Lansdowne, *Petty Papers* (London, 1927); *The double-bottom or twin hulled ship of Sir W. Petty* (Oxford, Roxburghe Club, 1931); Paul Hoste, *Théorie de la Construction des Vaisseaux* (Lyons, 1697); Pepys, *Naval Minutes* (N.R.S. 1925), pp. 158, 373.

[3] Claude Perrault, *Ordonnance des cinq espèces de colonnes selon la méthode des anciens* (Paris, 1683). Both Wren and Vauban were mathematicians before they were architects.

[4] Marin Mersenne, *Harmonie Universelle, contenant la théorie et la pratique de la musique* (Paris, 1636-7).

the evenings of his apprentice days 'to good purpose, as cyphering, drawing, and practising to attain the knowledge of my profession' and who later received lessons from John Goodwin, professor of the mathematics; or Samuel Pepys, at twenty-nine engaging the mate of the *Royal Charles* to instruct him in mathematics beginning with the multiplication table.[1]

The most obvious instance of this in England was the grant by Charles II in 1676 of a second charter to Christ's Hospital in order that this institution might become in part a technical school. Forty 'mathematical children', whose curriculum received the careful attention of Newton in 1694, were to be brought up for the sea-service at a cost of £370 10*s*. per annum met by the Exchequer.[2] Newton on this occasion recommended a sound practical training during apprenticeship founded on a firm theoretical grounding; experience is valuable, but he pointed out that there is as much difference between a mere practical mechanic and an educated one as between a rule-of-thumb surveyor and a competent geometer, or between an empiric in medicine and a learned physician.[3]

A similar trend towards technical education on the continent was even more marked; among its products were the military academies which, founded towards the end of this century, flourished particularly in the next. The French naval schools were the creation of the wars of Louis XIV.[4] In them, as in England, specialised training began with the teaching of geometry as an introduction to navigation or fortification, which since the advent of siege-guns had been reduced to regular rules by which ramparts, bastions, curtains and out-works were constructed in a polygonal figure with carefully calculated angles of inclination to the fire of the enemy. A lofty standard of competence was not to be expected, but educational theorists stressed the value of mathematics as a mental discipline, for teaching concentration, reasoning

[1] Pepys, *Diary*, I, 4 July 1662. Pett, *Autobiography* (N.R.S. 1918), pp. 7, 14.

[2] R. Ackermann, *History of the Colleges . . . and Free Schools* (London, 1819), Christ's Hospital, p. 9.

[3] J. Edleston, *Correspondence of Sir Isaac Newton and Professor Cotes* (1850), pp. 279 *et seq.*; L. T. More, *Newton* (1934), p. 402. Robert Hooke also was a Governor of the school.

[4] F. B. Artz, *op. cit.*; B. Poten, 'Geschichte des Militär Erziehungs und Bildungswesens in den Landen deutscher Zunge'. *Monumenta Germaniae Paedagogica*, X *et seq.* (Berlin, 1889) La Roncière *op. cit.* vol. VI pp. 86–7.

and the distinction between truth and specious appearance, in addition to the utility of simple sums and figures. The trend towards enthroning scientific thinking as the highest form of intellectual activity, in place of theological or philosophical speculation and linguistic exercises, was beginning. Even the Dauphin of France was taught fortification and machine-drawing by Bossuet.[1] Though the practical methods taught in the schools and textbooks were often framed in the light of rough and ready experience, making little or perhaps mistaken use of mathematical principles, though in fact the pupil had little guidance in his first steps beyond the lore of some illiterate but skilful craftsman, mate, or sergeant, the very fact that the attempt was made to teach techniques by mathematics, to substitute calculation for the practised eye of the master, marks the commencement of a new era.[2]

Gunnery, as we have seen already, was one of the crafts which received a veneer of mathematics, as a result of progress made at very different social and intellectual levels, and of a common interest bringing mathematicians and the practical users of artillery (or their mentors) to attack the same fundamental problems, whether they were embraced in the theory of projectiles, the art of gunnery, or the science of ballistics. Despite the longevity of traditional, cumbersome methods of manufacturing and using artillery, whose design remained for so long unaltered, military writers proclaimed scientific methods and standards through every decade, urging logical classification, standardisation, precise measurement and calculation.[3]

Approaching similar problems from the opposite direction, mathematicians found the problems involved in the calculation of trajectories of absorbing interest, especially from about 1670, when the difficulty of the operation equivalent to differentiation was being resolved. Many of the great mathematical problems successfully overcome in the seventeenth century were of this

[1] For the failure to uproot traditional methods cf. Artz, *op. cit.* p. 50. Charles II of England on at least one occasion expressed an interest in fortification (H. W. Turnbull, *James Gregory Tercentenary Memorial Volume* (London, 1935), p. 55).

[2] A corollary to this was the insistence that constructors should follow exactly the plans prepared which became noticeable in both the English Admiralty and the French War Ministry at this time.

[3] See above, p. 32.

type, arising from the application of geometrical methods to mechanics, as to the four conic sections were added successive groups of 'mechanical' curves, each presenting greater difficulties than the last.[1] Among the achievements of the century may be included the rectification of the cycloid, the calculation of centres of oscillation, the solution of the brachistochrone and not least the analysis of the trajectory of a projectile in a resisting medium. Although in Pierre Fermat France had one of the most active investigators of the theory of numbers, his tastes were not heartily shared by most of his contemporaries.

Further, the smattering of military knowledge that was normally a part of a gentle education and was usually to be found in a set course of mathematics gave additional point to the philosophical controversies over problems of motion, an example of the drawing together of the worlds of learning and practical affairs.[2] As an amplification of doubts concerning the new system of dynamics founded by Galileo (whom, however, he very much admired), Marin Mersenne (1588–1648), whose varied interests in science, philosophy and religion revealed in many works and an extensive correspondence gave him a leading place in the new intellectual movement in France, published in 1644 his *Ballistica et Acontismologia*, which popularised a new word in the European languages. Ballistics became, after navigation, the most obvious example of the application of mathematics to utilitarian purposes.

Thus we find such dissimilar writers as Christiaan Huygens and

[1] Cf. Bonaventura Cavalieri to Galileo, 3 Dec. 1630. 'Mi piace sommamente che habbi ripigliato le speculatione del moto, materia invero degna d'un par suo e che mi da straordinariamente nell' humore, vedendo che con tal scienza e con la matematiche accopiate insieme ci potiamo presentare alla speculatione delle cose naturali, e con gran confidenza sperarne la desiderate cognitione' (Galileo, *Opere*, Edizione Nazionale, vol. XIV, p. 171).

[2] Galileo taught fortification and wrote two treatises on it as texts for his pupils (*ed. cit.* vol. II, pp. 15–146). John Evelyn, like Descartes, played the soldier for a time in the Netherlands wars (*Diary*, 2–8 Aug. 1641). Otto von Guericke, inventor of the Magdeburg spheres, served under Gustavus Adolphus. Robert Boyle wrote an essay on fortification in his youth (*Usefulness of Mathematics*, 2). Torricelli probably taught the art of fortification (*Opere* [1919], vol. I, pp. ix, x). These are random instances. It is perhaps worth noting that Evelyn, in a classification of arts and crafts drawn up by him in 1660, places the gunfounder along with 'useful and partly mechanical' tradesmen, the gunner under 'polite and more liberal' (Royal Society, *Classified Papers*, vol. III, [1], no. 1). Examples of the mathematical compendium are *Jonas Moore's Arithmetic* (London, 1650), C. F. M. de Challes, *Cursus seu Mundus Mathematicus* (Lugduni, 1674).

Sir William Petty, the father of English economists, enumerating ballistics among the branches of mechanical science worthy of study.[1] It aroused the interest of Boyle, who in 1671 defined mechanics as including

those disciplines that consist of the application of pure mathematics to produce or modify motion in inferior bodies, so that in this sense they compromise not only the vulgar statics but divers other disciplines such as the centrobarricks, hydraulics, pneumatics and ballistics etc.[2]

He assumed indeed that such mathematical investigations were matters of common knowledge.

I care not to mention to you, (Pyrophilus) [he wrote] how great a variety of trials and observations about the best way of levelling great guns and the differing distances to which they will carry at such and such elevations, and the lines described by the motion of the bullet and other particulars belonging to the art of gunnery, have been proposed and tried upon the hints suggested by geometry's mathematical disciples (especially) and others because many good man with these fatal arts have been less understood.[3]

Leibniz, apparently, recognised that the utilitarian advantages to be expected from the encouragement of science were most likely to create favour at court, and included the study of military architecture and artillery with the improvement of manufactures as subjects fit for the consideration of a scientific academy.[4] François Blondel's *Art de Jetter les Bombes* was in fact published at the expense of the Parisian Académie des Sciences.

When scientific research and education were supported by royal patronage, as happened in most European countries before 1700, and the general welfare became a respectable motive for the study of physical science, public usefulness was not always clearly distinguished from militant patriotism, even in the minds of a Newton or a Leibniz. Few voices in the seventeenth century were raised against the use of war as an instrument of policy, and few

[1] Huygens, *Œuvres*, vol. XIX, pp. 25, etc.; *Petty Papers*, vol. II, p. 10.

[2] *The Usefulness of Mechanical Disciplines to Natural Philosophy*, p. 1. This is the first use of the word in English I have noted. The *O.E.D.*'s illustration is from the eighteenth century.

[3] *Usefulness of Mathematics*, p. 17.

[4] Foucher de Careil, *Œuvres de Leibniz publiées pour la première fois* (Paris, 1859–65), vol. VII, p. 317.

would have ventured to despise or condemn a new contribution to the art of destruction, at a time when it was a commonplace to link gunpowder, the compass and printing as the most beneficial inventions of the recent age, and to believe that the former had diminished rather than augmented the horrors of war. The pacifism of the Quakers was looked upon as but another instance of their perverted fanaticism, while natural philosophers entered without qualms upon the dangerous road of service to the state.[1]

Impetus and Galilean Ballistics

The position of Galileo in the development of dynamics has undergone a subtle change as the result of painstaking studies pursued in the last half century. Though no one will dispute that he was the first natural philosopher to work out correct mathematical concepts of inertia and acceleration, for instance, from which he was able to settle the ancient problem of defining the path described by a projected body, it has been recognised that, like other great men, he saw farther by standing on the shoulders of giants. He was the last and greatest figure in a medieval tradition of philosophy extending over three centuries; and the Galilean revolution in science was only possible because Aristotelian ideas were no longer deeply rooted in the mind as the inevitable starting-point for reason, but had been shaken by the theory of *impetus*. The advance of the science of moving bodies from its rudimentary classical level was caused less by a wider vision of phenomena than by the viewing of familiar appearances in the new and unprejudiced light that the philosophers of impetus had shone, without which the great feat of re-formulation and mathematisation effected by Galileo would have been impossible. His achievements should not be measured by comparing his science of dynamics with the absurdities of Aristotle, as he seems to suggest in his *Discourses on two New Sciences*, but by appraising its improvement on the revolutionary, already half-geometrised, notions of his medieval precursors.

[1] R. F. Jones, *Ancients and Moderns* (1936), pp. 2, 3, 13, 19, etc.; *Histoire de l'Académie Royale des Sciences* (1707), p. 121; W. C. Braithwaite, *The Beginnings of Quakerism* (London, 1919), pp. 462, 519-22. Boyle has a noble passage on the merits of healing in contrast with destructive arts in the *Usefulness of Physic*.

The introduction of the term *impetus* into philosophy was brought about by a revolt against the Aristotelian theory of projectiles. Originally only a minor variant in the whole corpus of peripatetic physics, as the idea was developed important corollaries followed from it. Aristotle had taken as the foundation of his argument the proposition that no inanimate body can move without a motive force; unless it is able to fall like a ripe apple, or rise like the smoke from a fire, the lifeless furniture of the world has no power of locomotion. But there was another simple and obvious phenomenon that seemed to conflict with this proposition, for an arrow will fly through the air or a ball roll along the turf or a grindstone turn, after the motive power of man has been removed, and the philosopher's problem was the reconciliation of this everyday experience with his fundamental axiom. What force moves the arrow in flight, what determines the distance it shall cover before it drops exhausted to the ground? Aristotle's own resolution of this difficulty was to introduce a new factor, the medium in which motion takes place. To the medium, be it air or water, he attributed diverse functions; in natural motion, which is that of a heavy body falling directly towards the centre of the earth, or a light body rising directly from it, the medium resists. This was a conclusion which no one who had ever watched the fall of a broad leaf from a tree could well escape. Violent motion, the opposite in all things of natural, being the upward or horizontal movement of a heavy body, or the downward movement of a light, and always the result of the use of force to overcome natural tendencies, is assisted by the medium. Indeed without the medium there could be no violent motion, and if a vacuum were possible it would be impossible to shoot an arrow through it.[1]

In rejecting Aristotle's explanation of continued motion by a faculty in the medium of storing moving force, Jean Buridan and Nicholas Oresme in the fourteenth century revived some ancient classical objections and by stating them more forcibly and exploring them further, they and their successors created a new

[1] *Physics*, 241 b, 254 b, 258 b; *De Caelo*, III 2. A vacuum is impossible because without the resistance of the medium natural motion would be instantaneous. (*Physics*, 215 a, *et seq.*)

dynamics.[1] The essence of the impetus theory was its denial that the medium had a dual function to resist natural and assist violent motion. The distinction between these two species of motion was preserved, but both were thought to be resisted. Quiescence or natural motion were still the normal states of a body, only, if it were once set in violent motion by some force, such as that of a bow, it would receive an evanescent tendency to continue in motion. The arrow does not at once fall to the ground because it has received an impetus proportional to the force impressed and to the weight of the body, just as a piece of iron withdrawn from the fire remains hot for a long time. Impetus was a quality of moving which a body might possess through being moved, as it might possess the quality of warmth through being heated.[2] For this reason it is possible to throw a heavy body like a stone further than a feather, because it can absorb more impetus.

These in principle were the doctrines which, mainly through Albert of Saxony, reached Leonardo da Vinci and directly or through his note-books became the common property of the sixteenth century, influencing Tartaglia and Cardan, Benedetti and Bonamico, who was Galileo's master.[3]

The same stream of ideas which brought the theory of impetus to the Italian universities of Galileo's youth carried with it a non-Aristotelian explanation of the descent of heavy bodies to the earth. No attempt was made to suggest then, or for a very long time, an alternative to Aristotle's reasons *why* they fall, but to give a philosophical account of *how* they fall. Buridan wrote that the rate of descent increases because a stone, for example, has, added to its natural propensity at any instant during the descent to begin falling, the impetus acquired during the fall up to that

[1] The history of the theory of impetus and the evolution from it of the classical laws of dynamics have been very fully treated by Emil Wohlwill in the *Zeitschrift fur Volkerpsychologie und Sprachwissenschaft*, vols. XIV, XV; Pierre Duhem in *Études sur Léonard de Vinci* (Paris, 1906–13), especially vol. III, pp. 34–54; and A. Koyré in *Études Galiléennes* (Paris, 1939). Reference should also be made to *Medieval Studies* (New York and London), vol. III, pp. 185 *et seq.* 'Maistre Nicole Oresme: Le Livre du Ciel et du Monde'.

[2] This illustration is used by Galileo in his early treatise on impetus dynamics *De Motu* (*Opere, ed. cit.* vol. I, p. 310). The impermanence of impetus is stated by Leonardo in terms that can only be described as mystical (MS. A, Institut de France, ed. Ravaisson-Mollien, 34 v; Codex Atlanticus 219 v. (a); J. P. Richter, *Literary Works of Leonardo da Vinci* (London, 1939), vol. II, p. 219).

[3] Duhem, *op. cit.* vol. III, pp. 181–213; Girolamo Cardano, *De Subtilitate Libri xxi.*

instant, which is continually augmented.[1] Oresme showed by a simple geometrical demonstration that the mean velocity of descent is equal to that attained at the half-way point.[2] Albert suggested that the instantaneous velocity must be proportional either to the time elapsed since the start of the accelerated motion or to the space passed over, and Leonardo favoured the second of these hypotheses.[3]

These and other writers who accepted the theory of impetus used it in their efforts to discover what sort of path a projectile follows through the air. From this distance of time it is easy to see that four major steps had to be taken, each demanding the most daring inspiration, before it was possible to study even the simplest trajectory: to imagine that the influence of air resistance had been removed; to resolve the motion of the projectile into its component parts; to unfold the idea of inertia which alone could make linear motion comprehensible; and to define acceleration. How far did the impetus theory go towards making these advances possible? It did suggest that uniform movement along a straight line was possible and it did provide a loose and inaccurate concept of acceleration; on the other hand it confused two effects which it was most important to separate, namely the slowing down of the projectile caused by air resistance, and its decline to the earth due to gravity. Owing to this confusion the theory suggested that a projectile was supported in the air by its impetus, then as this relaxed, fell to the ground. Accordingly Oresme divided the flight of a projectile into three portions, in the first of which it moves in contact with the propellant, its velocity increasing; in the second it leaves the propellant and its velocity goes on increasing but the 'generation' decreases, in the third the natural movement towards the earth prevails. Thus he accounted for the 'fact', still half-believed in the middle of the seventeenth century, that a projectile has a more violent impact in the middle of its path than at the beginning or end.[4]

Albert of Saxony regarded the path of a missile as the product of three mutually opposing forces, the impetus, the resistance, and

[1] Duhem *op.cit.* vol III, p. 41.
[2] *Ibid.* p. 395.
[3] *Ibid.* pp. 512 *et seq.*; Leonardo's MS. M 43 r., 44 v. Cf. Paul Tannery, *Mémoires Scientifiques*, vol. XIV, p. 224.
[4] *Medieval Studies*, vol. IV, pp. 225–6.

gravity. The first part of the trajectory is described when the impetus is greater than both gravity and resistance, it becomes curved when the impetus is greater than the excess of gravity over resistance, and finally when gravity overcomes both resistance and the failing impetus, the body falls to the ground.[1] This analysis was adopted with modifications by Leonardo and thus rendered popular in the sixteenth century. The picture of the trajectory which he drew, borrowed by Tartaglia and Cardan, appears almost unchanged in manuals of gunnery to the end of the next. The first and last parts of the path of the bullet are straight lines, the former because impetus or violent motion prevails in it, the latter because gravity or natural motion prevails. Men continued to debate whether, as in Leonardo's view, a real 'mixture' of the two types of motion produced the curvature about the vertex.[2] Thus when a ball is shot from a gun at an angle to the horizon, it rises in a straight line to a certain height proportionate to the charge of powder, then as the impetus weakens moves in a curved path and finally finishes its descent in a straight line towards the centre of the earth. Unless the decreasing density of the air in the upper regions (where it mingles with the lighter element of fire) must be taken into account, the height and the range, resembling the altitude and base of a right-angled triangle, will be strictly proportional to the strength of the charge of powder.[3]

The same reasoning was used by Cardano in *De Proportionibus*.[4] In the hands of Tartaglia it was modified by his perception that gravity, however violent the impetus, must prevent the projectile following a perfectly straight line.[5] Daniel Santbech applied geometry to this same trajectory as originally conceived by Leonardo on the ground that although the path of the shot becomes curved at *B*, it may be treated mathematically as though the shot continued along the hypotenuse to *C* and then fell vertically to *D*. Accordingly the ballistical triangles formed by shooting at the same angle with different charges, *Acd* and *ACD*, are similar and the range *Ad* is in simple proportion to the range

[1] Duhem, *op. cit.* vol. II, pp. 215–16.

[2] MS. A, 1 v., 4 r., 35 r., 43 v., etc.

[3] MS. M, 53 r.

[4] Bk. V, Prop. XC.

[5] *Nova Scientia*, bk. II, Supposition II; *Quesiti et Inventioni*, bk. III, pp. 11–12. Leonardo seems to have been the first to draw the trajectory as a continuous curve in his illustrations.

AD; moreover, since the hypotenuse *AC* is equal to any other *AQ* described by a projectile propelled with the same force, the ranges at different angles with the same charge of powder in the gun, are as the cosines of the angles of elevation.[1]

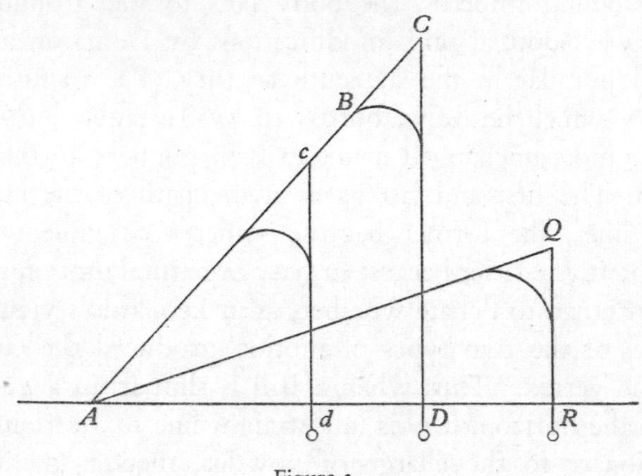

Figure 3

Benedetti on the other hand avoided such extrapolations of Leonardo's original theory, describing the path of a projectile as produced by the mixture of natural and violent motions, though either impetus or gravity alone would produce a rectilinear motion.[2]

Such briefly were the suggestions that had been offered towards the solution of the ballistic problem that Galileo was to make peculiarly his own, and though he was not altogether in the position of owing his predecessors no debt of gratitude, the slight praise he awarded to them, the contrast he pointed between their vague and hypothetical declarations and his own precise demonstrations of the true principles of dynamics hardly exaggerated the extent of his renunciation of the impetus mechanics which had been the cradle of his first thoughts. The theory of impetus, which had begun so auspiciously with a doubt of Aristotle's omniscience and an appeal to observation, had failed to establish itself as an independent branch of natural philosophy, and was buried in grand explanations of the whole universe. It provided a novel

[1] *Problematum astronomicorum et geometricorum sectiones septem* (Basileae, 1561), Props. 114–18.
[2] Duhem, *op. cit.* vol. III, p. 218.

explanation of the continuation of movement or the acceleration of falling bodies, but it did not in spite of some few attempts, bring out the importance of discovering how the movements of bodies may be calculated. It lent itself to mathematical description no better than did Aristotle's theory, and it caused no disruption of the whole scheme of his physics. It was simply absorbed. Even the invention of artillery did not lead to a fresh investigation, for the early theories of gunnery (lasting well into the second half of the seventeenth century), as we have seen, accepted impetus physics and Albert of Saxony's analysis of the trajectory in every detail. Mach was, of course, wrong in supposing that statics was the whole of mechanical science before the Galilean revolution, but it is true that dynamics as a science in the modern sense was founded by Galileo.[1]

The twin pillars of the new framework of nature took shape in his mind early in the new century.[2] The relations of time, space and velocity in the fall of heavy bodies were already known to him in 1604, though as yet he could not give a rigorous demonstration of them.[3] He had turned his attention to the study of the theory of projectiles in the light of his new ideas by 1609, when he wrote that he had made the discovery that if a piece of artillery is taken to some height above a plain and fired with different charges, the bore always being horizontal, though the ranges will be different the shot will always graze the earth at the end of the same number of seconds because the velocity of the vertical motion is unchanged; further, the time of flight will be constant for very different ranges, no matter what the elevation, if the vertical height above the ground reached by the bullet is the same.[4] He did not publish anything of this until his researches were complete, but in his *Dialogo sopra i due Massimi Sistemi del Mondo*, printed in 1632, Galileo devoted a number of pages to the explanation of his rule, here first made known to the world, that when a body falls the spaces passed over in equal time intervals

[1] Ernst Mach, *The Science of Mechanics* (Chicago, 1907), p. 128.

[2] For the genesis of Galileo's ideas, see Koyré, *op. cit.* vol. III.

[3] Letter to Paolo Sarpi, 16 Oct. 1604 (*Opere*, vol. X, p. 115). The first demonstration on exclusively mathematical principles was that of Huygens (*aet.* 17) (*Œuvres*, vol. I, p. 27).

[4] To Antonio de Medici, 11 Feb. 1609 (*Opere*, vol. X, p. 229).

are as the series of odd numbers 1, 3, 5, 7, etc., and of the law of inertia.[1]

He also discussed the problem of the real path in space of a stone falling from a high tower, which on the Copernican hypothesis could not be a straight line, though relative to the earth it appears to be so. The stone is really a projectile in space, since its motion is the resultant of two primary motions, the one towards the centre of the earth, the other an inertial rotary motion (in Galileo's view) derived from the earth's circular motion, 'cal quale composto ne risalterrebbe che'l sasso descriverrebbe non piu quelle simplice linea retta e perpendicolare ma una trasversale, enforse non retta'.[2] However, he did not take the opportunity which this topic might have offered of proving that the path of a projectile is a parabola, if indeed he was aware of this at the time, for he believed (and went on to demonstrate) that the path of the falling stone 'movendosi del moto del commune circolare e del suo proprio retto' was a semi-circle having a radius of the earth through the point of origin as diameter.[3]

This is one of the rare instances of Galileo's lapsing into error, in this case because his conception of the law of inertia was imperfect. It is curious that the chief merit of Galileo's exposition of inertia, his grasp of the relativity of all local motions at the earth's surface, should have led him, while giving so many hints towards the solution of the central problem of ballistics, to lose the priority of publication and to give a wrong answer to an analogous question. The honour of being the first to demonstrate that a projectile follows a parabolic trajectory (if the effects of air resistance are neglected) fell to Bonaventura Cavalieri, a mathematician rather than a physicist, whose book on *The Burning Glass* came out later in this year 1632.[4] He had been working on this treatise, with others, including the famous *Geometria indivisibilium continuorum*, since 1629, but it was only in August 1632, when the printing of the book was already complete, that he wrote informing Galileo of a passage in it dealing with the motion

[1] *Opere*, vol. VII, pp. 163 *et seq.*, 190 *et seq.*, 248 *et seq.*

[2] *Ibid.* p. 165.

[3] *Ibid.* pp. 190–1.

[4] *Lo Specchio Ustorio overe Trattato delle Settioni Coniche, et alcuni loro mirabili effeti intorno al lume, caldo, freddo, suono e moto* (Bologna, 1632); Galileo, *Opere*, *ed. cit.* vol. XIV, pp. 59, 354, 377–8.

of projectiles on the principles of mechanics laid down by him in the first series of *Dialogues*.

In chapter XXXVIII of *Lo Specchio Ustorio* Cavalieri explains how two forces, that of the propellant and that of gravity, act upon the projectile, each of which acting alone would cause it to follow a rectilinear path. If each of these combined forces separately impelled the projectile uniformly in different directions, so that it passed over equal spaces in equal times, the resultant trajectory would still be a straight line at an angle to the direction of propulsion, but because, as Galileo had shown in his *Dialogues*, the force of gravity causes a body to accelerate and pass over distances which are proportional to the square of the time of fall, the path is not a straight line but some sort of curve. In the next chapter Cavalieri plots this curve between rectangular coordinates, calculating the horizontal and vertical distances traversed in successive equal time intervals, and proves that it is a parabola. He admits that the trajectory should really end at the centre of the earth and that his vertical lines are not truly parallel, but only approximately so for the distances involved in shooting, but of more importance than these approximations when ranges are long is the resistance of the air, which opposes both the impulse and gravity. Even this he proposed to disregard because it is less obvious over a short space and cannot be easily assessed or dealt with in a few words.[1]

The master, however, was not pleased with the work of his disciple in constructing the theory of projectiles on his own principles and in reaching a conclusion that he looked upon as exclusively his own. He wrote to Cesare Marsili, another Bolognese, that he had heard of Cavalieri's newly published book and of the proposition therein inserted that a projectile moves along a parabolic path.

I cannot conceal [he declared], that such news was of little relish to me, to see how the first fruits of my study over more than forty years, imparted with great trust to that Father [Cavalieri was a Jesuit] were now stolen from me, and the renown which I so keenly desired and had promised myself from my long labours, tarnished; for truly my first purpose, which moved me to speculate on motion, was the

[1] *Op. cit.* pp. 168–9.

discovery of such a line (which most certainly when found is of little difficulty in demonstration); nevertheless I, who have proved it, know what pains I have had in discovering that conclusion.

If Cavalieri had communicated his thoughts to Galileo before the publication of his book, as true politeness required, he would have begged him to defer it until his own work had appeared, after which Cavalieri could have added as many discoveries as he pleased.[1]

This letter of lamentation over the injustice done him by Cavalieri would perhaps hardly be worth quoting did it not demonstrate so strikingly the importance which Galileo attached to the culmination of his discoveries in mechanics, the proof that the trajectory of a projectile was a precise, definable, even familiar geometric curve. Galileo, like Newton, learnt, while never admitting, the unfortunate truth that there is no patent in scientific ideas until they have been firmly displayed in public. So far as can be discovered, though it may be true that the application of the laws of motion to the solution of the trajectory was obvious to even a mediocre mathematician after the publication of the *Dialogues*, Galileo himself had not communicated a proof of what he claimed as his own discovery before 1632.

It is interesting to compare Cavalieri's extension of Galileo's laws to their proper fulfilment with the commentary of Mersenne on the passage of the *Dialogues* already mentioned, where Galileo decided that the path of a falling stone is really a semi-circle. Mersenne was able to show that, since radii of the earth separated by only a short distance at its surface are approximately parallel, on Galileo's own principles the path of the stone must be very nearly parabolical.[2] Thus there is no need to suppose Cavalieri guilty of plagiarism because he had printed a theorem which, in a modified and less general form, was reached independently by Mersenne, who was certainly no great mathematicion. In any case the ill-feeling between him and Galileo did not last long. Cavalieri was apologetic even to the extent of offering to withdraw his book, and lesser troubles were obscured as the shadow of ecclesiastical displeasure fell over Galileo.

[1] Galileo to Cesare Marsili, 11 Sept. 1632 (*Opere*, vol. XIV, p. 386).

[2] *Harmonie Universelle. Livre des Mouvemens de toutes sortes de Corps*, pp. 93-8.

The further progress of the theory of projectiles in the seventeenth century, indeed all the future development of exterior ballistics, has seen the closer approximation of the parabolic trajectory to physical observations. But the statement made by Cavalieri was rudimentary and incomplete; the amplification of the new knowledge, the examination of the consequences that followed from it, was the work of Galileo himself and of Evangelista Torricelli, his last and favourite pupil.

It has already been mentioned that Galileo wrote on motion in the traditional impetus language about 1590 and that in 1609 he was occupied with reflection and experiments ('diverse esperienze') towards his treatise on mechanics, in which he already promised to bring forth new principles.[1] These studies came to fruition in the *Discourses on two New Sciences* published in 1638. It was never Galileo's method merely to expose the truths of science as he conceived them, but always en route in his reasoning to dispose of earlier theories and override all possible objections to his own. In this case he asserted his own doctrines by attacking with the greatest vigour the Aristotelian doctrines of the schools. As a result the *Discourses*, though less turgid than the *Dialogues*, are unbalanced between the unfolding of new ideas, historically the most significant portion of the work, and the refutation of old ones whose falsity only now seems self-evident. Therefore in the progress of his attack on the peripatetics the unity of a study primarily devoted to the science of motion is broken by an essay on the strength of materials. The well-known argument on falling bodies occupies the First Day: the theory of projectiles which logically follows is deferred to the Fourth.

The proof of Galileo's cardinal First Theorem is too well known to require repetition and differs very little from that of Cavalieri. For justification of his assumption that the radii from points on the trajectory to the centre of the earth are parallel Galileo appealed to the example of Archimedes who had made a similar supposition in his writing on the balance; even if the distance between the point of origin and the point of impact is as much as four miles, this distance is inconsiderable in relation to the diameter of the earth.[2] A more serious objection that the path

[1] See above, p. 85, n. 4.

[2] *Opere*, vol. VIII, pp. 274–5.

traced by a uniform horizontal and a downward accelerated motion is not necessarily the same as that traced when the uniform motion is oblique was only disposed of in a passage drafted after the publication of the first edition of the *Discourses*, in which Galileo showed that, whatever the angle of projection, the trajectory is always a greater or less portion of a parabola.[1]

The first concept to establish in the exploration of the parabolic trajectory was the measurement of impetus, for he continued to use the word, though it no longer explained all dynamical phenomena. To Galileo impetus was equivalent neither to acceleration nor to uniform velocity, but something which might produce either of these. Accordingly he used a method of measuring impetus, velocity and acceleration together by means of his rule for falling bodies. Imagine a weight to fall a distance represented by a line of given length. The time taken, and the velocity of the body at the end of the line, are known from the rule, while its impetus is just sufficient to carry it back to its starting-point if reflected upwards over the same path. Conversely, all these if known can be laid down as a straight line. Although this method seems cumbersome in comparison with the modern statement of acceleration as so many units per second squared, it was extremely well adapted to the seventeenth-century geometrical mathematics, since it provided a means of measuring a rate by a line. By this means, Galileo proved what had long been suspected, that the amplitudes of parabolas, in practice the ranges of shots, are equal when the angles of projection at a given velocity are complementary. He also showed that the amplitude is proportional to the altitude and therefore to the velocity, giving constructions for finding the third of these variables from the two known.

So far the discussion of ballistics by Galileo was entirely theoretical, even academic, whereas Cavalieri had made plain in the proof of his proposition the sort of projectile he was considering. Only two of the heads for a military treatise drawn up in earlier years by Galileo had been touched upon, and without relation to practical circumstances.[2] But the range tables printed as an appendix to the *Discourses* could have none but a practical interest, nor need it be supposed that their compiler had any doubts of

[1] *Opere*, p. 446.

[2] Cf. *Opere*. vol. VIII, p. 424.

their value as a guide to gunnery. In many passages Galileo remarks that the theory of projectiles is of great importance to gunners. He made little or no distinction between his theory and useful ballistics; he believed—though without experiment—that he had discovered methods sufficiently accurate within the limitations of military weapons to be capable of direct application in the handling of artillery. There can be no doubt that, if the level of the argument of the book was beyond the attainment of the gunner, parts of it were written for the professional experts on the art of war, who in fact did in course of time accept Galileo's straightforward proofs and simple conclusions (without the necessary provisos of which he was well aware) and embalmed them in the manuals of training. Criticism of Galilean ballistics came came not from soldiers but from scientists.

His extensive analysis of the parabolic trajectory by geometric methods contains no general equation for the curve, nor the simple expression relating the ranges of shots at different angles of projection. Galileo had calculated each range in his tables from a chain of proportionals. Examining these calculations, Torricelli was able to demonstrate that from them it followed $r = 2R \sin \alpha \cos \alpha$; or $r = R \sin 2\alpha$, where r is the range at any angle α and R the extreme range. Thus if the range at 45° elevation is found to be 1,000 paces the ranges of 34·9 paces at 1°, 69·8 at 2° and so on can be read off from an ordinary table of natural sines, of which in effect Galileo's table was a portion. This discovery, rendering the construction of any range table a very simple matter, was published in an essay in which he took up and completed the work of his former master.[1]

Torricelli declared that the doctrine of projectiles was the supreme result of Galileo's efforts, as likewise his supreme glory.[2] While, therefore, primarily inspired by veneration of Galileo, Torricelli is more logical in his treatise and bears the application

[1] *Opera Geometrica* (1644), *De Motu Gravium . . . in quibus ingenium naturae circa parabolicam lineam ludentis per motum ostenditur, et universae projectorum doctrina unius descriptione semicirculi, absolvitur*, bk. II, 'De motu projectorum'. Torricelli had an interest in ballistics from the time of his first acquaintance with Galileo's works. In 1641 he discovered that the envelope or curve touching all the trajectories described from one point with a given velocity of projection is itself a parabola whose latus rectum is twice the greatest range (*Opere* [1919], vol. III, pp. 43, 130).

[2] *Ibid.* p. 154.

of his work constantly in mind. He defines the direction of projection as the tangent to the trajectory at its point of origin, which is also in artillery the axis of the gun, and proves by Galileo's law of acceleration that the greatest altitude of an ascending body is half that height which it would have reached if it had ascended with a uniform instead of a decelerating vertical velocity. His proof that the path of a projectile is parabolic is more direct than those of Cavalieri and Galileo. If *ACDEF* is the trajectory, its vertex at *F*, and *AB* the direction of projection, from the relation already found $BF=FG$. Let *AB* be divided into a

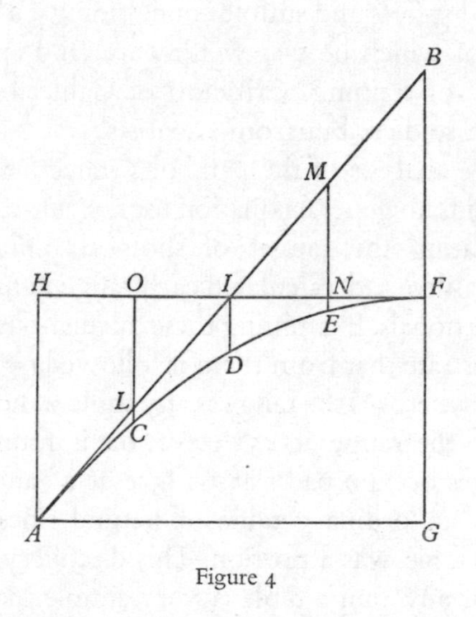

Figure 4

number of equal parts *AL*, *LI*, *IM*, *MB* and the perpendiculars be drawn to the curve; these equal spaces are passed over in equal times during which the body also falls the distance *LC*, *ID*, *ME*, *BF*, which are to each other by Galileo's rule as 1, 4, 9, 16.... The divisions *HO*, *OI*, *IN*, etc, are also by construction equal, and, since $OL=MN=\frac{1}{2}BF$, by subtraction the distances *NE*, *ID*, *OC*, *HA* are as 1, 4, 9, 16. Therefore the curve *ACDEF* is a parabola.[1] In enumerating fresh properties of the trajectory, Torricelli shows that the force (impetus) required to carry the projectile from *A* to *F* is larger than that needed to bring it down from *F* to *A* in the ratio of *AB* to *AG*; that the impetus of the

[1] *Opere*, vol. III, pp. 157 *et seq.*

projectile at any point on the curve is in proportion to the length of the tangent at the point intercepted between two perpendiculars; and that the impetus is the same at corresponding points on the ascending and descending branches of the curve.[1]

Maintaining the geometrical analysis of Galileo, Torricelli much improved the use of a line which Galileo had called the 'sublimate', that is, the difference between the maximum height attainable by

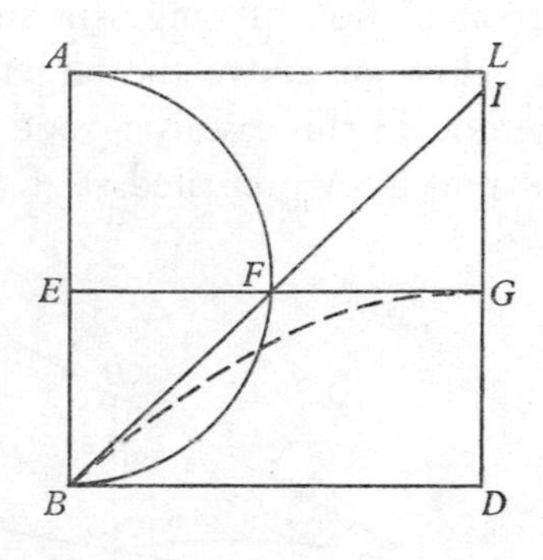

Figure 5

the projectile $\frac{V_o^2 \sin^2 \alpha}{2g}$ and its actual height. To construct the parabola described with a given velocity at a given angle he evolved the following procedure: suppose *AB* to represent the distance which a body would fall freely to acquire the velocity *V*, and *DBI* the angle of elevation. To find the amplitude and altitude of the parabola described by a projectile having the initial velocity *V*, inscribe the semicircle *AFB* on *AB* to which *BD* (the horizontal) and *AL* are tangents parallel to each other. If *BI* cuts the circle in *F*, draw *EFG* parallel to *BD* so that *EF*=*FG* and complete the parallelogram *BALD*. Then *DG* is the diameter of the required semiparabola, and *LG* the sublimate representing the velocity at the vertex *G*.[2]

This way of constructing the parabola, incidentally, brings out several of the properties of the trajectory; for instance, it is easy

[1] *Ibid.* pp. 161 *et seq.*

[2] *Opere*, vol. III, pp. 165-6.

$$AB = \frac{V^2}{2g}. \qquad BI = 2AB\cos(90° - \alpha) = 2AB\sin\alpha.$$

$$\therefore\ DG = \tfrac{1}{2}ID = \tfrac{1}{2}BI\sin\alpha = AB\sin^2\alpha = \frac{V^2\sin^2\alpha}{2g}$$

when the point F is found to measure the height and range, and to see that the range of a shot ($4=EF$) is greatest when it is fired at 45° Thus this construction based on a semicircle was adapted in various ways to make it suitable for gunner's quadrants from which after calibration the ranges at different angles of fire could be read directly.

Torricelli also gave solutions for a problem which Galileo had not attempted, the calculation of ranges in such situations that the gun lies either below or above the target or plane of fire. His geometrical method in this case is a good deal simpler than the trigonometrical proof. A gun sited at *A* firing at the angle

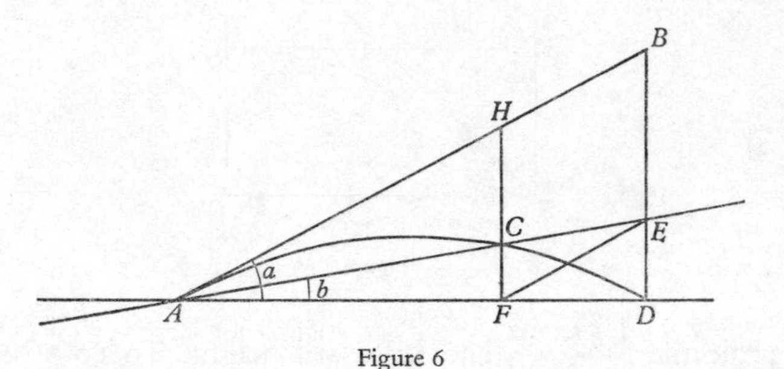

Figure 6

DAB has the horizontal range *AD* and the distance that it will shoot over the sloping plane *AE* is sought. Erect *BD* perpendicular to *AD* meeting the line of projection at *B*, and draw *EF* parallel to *AB* and *HF* perpendicular cutting the plane *AE* at *C*. Then *AC* is the range of the gun.[1] A similar problem occurs when the gun is, as usual, mounted on a carriage or fortification above the level over which it fires, for then the elevations necessary for its ranges will be very different from those found in the tables. The point-blank range especially will be very much increased. Torricelli found that under these conditions the point-blank range is equal to the square root of twice the height of the gun multiplied by the extreme range.[2] If it is elevated its range

[1] *Opere*, vol. III, pp. 188, 221. Simple trigonometrical considerations show that

$$AC=\frac{AD}{\cos b}\left(1-\frac{\tan b}{\tan a}\right)=\frac{2V^2\cos a\sin(a-b)}{g\cos^2 b}.$$

[2] *Ibid.* p. 223. If the extreme range is R the velocity of projection is $\sqrt{(Rg)}$. The time taken by the shot to reach the ground is $\sqrt{\left(\frac{2h}{g}\right)}$, therefore the distance passed over is $\sqrt{(2hr)}$.

will become

$$\tfrac{1}{2}R\left(1+\sqrt{\left(\frac{a+h}{a}\right)}\right)$$

where R is the normal horizontal range of the gun at this angle, a the normal altitude of the trajectory (which data may be found from the tables) and h the height of the gun.[1]

Torricelli did not imagine that his theory, presented in all its mathematical complexity, was likely to find much practical application in the hands of gunners. He remarks that with the aid of the quadrant invented by Tartaglia some of them had acquired considerable expertness in shooting at ranges beyond the point-blank and had learnt what elevation to give a culverin of forty pounds ball, for example, to strike a mark at a distance of 700 paces.[2] But there was little exactness in their methods, and it was necessary to repeat the experiments with every type and weight of gun and every strength of powder; moreover, it was utterly mistaken to suppose that the ranges increased in simple proportion to the increase in elevation. There was no way of reducing their observations to a rule without the aid of theory and geometry, based upon the proposition of Galileo, who first discovered and taught that a projectile moves in a parabolic line.[3] To make his discoveries more accessible to the unlearned, therefore, Torricelli published tables of ranges and altitudes from which, the range at any one angle having been measured, the rest could be found by the rule of three. To simplify the matter still further, he described the construction of an improved quadrant to serve the same end without calculation which, even if the parabola became very deformed through the resistance of the medium, or the ranges turned out to be irregular through many other chance occurrences, would undoubtedly satisfy the school of mathematicians if not the bombardiers.[4]

This work has been described at some length because it represents a full statement of the simplest ballistic theory, which was taught throughout the rest of the seventeenth century and well into the eighteenth. Torricelli had accurately stated solutions of the principal problems which would confront the mathematically

[1] *Opere*, vol. III, pp. 224–5. [2] *Ibid.* pp. 226–8. [3] *Ibid.* pp. 228–9.
[4] *Opere*, vol. III, p. 229. The tables are to be found on pp. 204–15.

minded gunner, and as he pointed out he had even included matters which were rather of scientific than of practical interest. Later writers followed Torricelli in every detail though their proofs and constructions might be different; and the classical geometric method of Torricelli's exposition was only displaced at the very end of the century. When Thomas Venn and John Lacey printed their book of military training in 1672, *Military and Maritine Discipline*, they thought fit to publish under the title 'The Compleat Gunner' translated extracts from 'The doctrine of projects by those late famous Italian authors Galilaeus and Torricello . . . that the benefit of his [sic] pains might redound to the English reader that is especially delighted or exercised in the affairs of Mars'.[1] So little had times changed. Blondel[2] and Halley[3] in their essays into the field of ballistics stated in different forms what was already known. The former indeed published designs by Cassini, Roemer, and De la Hire for an improved gunner's quadrant which made possible the allowance for the slope of the ground in assigning angles of elevation, and the latter expressed his results in algebraic form; but Blondel's instruments were never used and Halley's formulae are too involved to be judged an improvement on Torricelli's geometrical constructions, though he must be noticed as the first to remark that the extreme range of a gun on the parabolic hypothesis is always obtained when the axis of the piece bisects the angle between the vertical and the ground, whether this slopes or not.[4]

Fortunately for the future of scientific inquiry Galilean ballistics were not at once accepted without challenge. The criticisms levelled by other philosophers will be discussed later: for the moment it will be sufficient to examine Torricelli's defence of his own results, which throws some light on the scientific mind of the period. Galilean ballistics were not and could not be the fruit of experimental method; experiment, even with slow-moving bodies, at once reveals that projectiles do not move in a parabola. Torricelli, when pressed on this point, always claimed that his study *De Motu Projectorum* was purely theoretical, that it explained

[1] *Op. cit.* Preface. [2] *L'Art de Jetter les Bombes* (1683).

[3] Halley's articles published in the *Philosophical Transactions of the Royal Society*, vols. XVI, XIX.

[4] *Phil. Trans.* no. 216, pp. 69-70.

not what is observed to happen but what would happen under specified conditions, and that his purpose was the mathematical consideration of such hypothetical motion. Torricelli demanded to be treated as a philosopher and a mathematician, not as a physicist. His attitude was not unnatural, indeed it was the only one he could well take up. Confronted with experimental contradictions to his theory, the pupil of Galileo was not to say, 'The experiments are wrong'; nor could he doubt that the reasoning from the premisses he had used was impeccable. As for these physical premisses—Galileo's laws of motion—they had that great authority and, if they were displaced, what was to succeed them? The only possible answer was to explain that the theory was, after all, a theory, and that if nature did not conform exactly, then results found by trial would be different from those expected.

Thus Torricelli's reply to Roberval, who had objected against the Galilean theory of motion that the air is a resisting medium and so there can never be either a uniform or a uniformly accelerated motion, reminded him that this very objection had already been discussed by Galileo, and asked him to discuss the theory solely on its mathematical merits. Archimedes believed that projectiles move in a spiral and wrote a whole book about that curve; was his geometry any the worse because his theory was wrong? Geometry to Torricelli was independent of space and matter—take away all talk of projectiles, heavy bodies and balistas and the geometry would remain in a series of abstract propositions. The tables and instruments he had described were not for measuring the ranges of cannon balls, but certain geometric lines associated with geometric parabolas.[1]

A complaint which was even more concrete and precise (and shows the danger of unbelief to which even the mechanical sciences were exposed) came from Giovan Battista Renieri, a Genoese who wrote that Torricelli's book having arrived in that city a group of gentlemen had become interested in making experiments in gunnery with several types of artillery 'and indeed I was astonished that such a well-founded theory responded so poorly in practice'. The point-blank ranges were much longer than had been expected.

[1] Roberval to Torricelli, 1 Jan. 1646; Torricelli to Roberval, 7 July 1646 (Torricelli, *Opere*, vol. III pp. 352–4, 384).

If the authority of Galileo, to which I must be partial, did not support me, I should not fail to have some doubts about the motion of projectiles, whether it is parabolical or not, and whether, if so, the axis of the parabola ought to be perpendicular to the horizon or not.[1]

In his reply Torricelli adopted a tone which has an appearance of simplicity and logic. He had written assuming the truth of two hypotheses, Galileo's laws of inertia and acceleration and there had been no intention of proving these. But the first must be true unless the moving body meets with something to add to or detract from its velocity; the second is a close enough approximation since nature must always observe the same law and no other law will fit so well as that of Galileo. If both these assumptions are accepted then all that he and Galileo had written must follow. As for discrepancies arising in practice, Galileo had discussed various reasons for these, of which the chief was the impediment of the air. Either everyone must agree with the hypothesis and all that followed from it, or else they must be proved false, and then he would be willing to abjure all his mechanics.[2] In a second letter he stated outright that his book was written for philosophers, not gunners.[3]

This argument would be very convenient for anyone who wished to show that Galilean ballistics had nothing to do with technology and were entirely speculative. According to Torricelli they had no connection with practical gunnery or with real projectiles. But it would be disingenuous not to point out that his answer is specious, though it was the only answer he could well make. Cavalieri, Galileo and Torricelli had each stated that he was neglecting the resistance of the air; the real issue is whether they had made it clear that this reservation completely vitiated the parabolic theory from the experimental point of view. It was not merely a question of minor failures in which observation did not agree with theory; but a doubt whether the theory itself must be totally inadequate to explain observation. Torricelli at least had failed to make this distinction. It was pardonable for the reader to suppose that when he talked of guns he meant real guns,

[1] Renieri to Torricelli, 2 August 1647 (*ibid.* pp. 459 *et seq.*).

[2] August 1647 (*ibid.* pp. 461 *et seq.*)

[3] 1 September 1647 (*ibid.* pp. 478 *et seq.*).

that when printing tables giving measurements in paces he was not merely computing lines on a diagram, that when he designed a gunner's quadrant it was not to be used in the geometrical barrels of hypothetical guns. Since Torricelli's book has all the apparatus of reality the reader had good cause for complaint if it was wrong. He could justly argue that Galilean ballistics, on Torricelli's showing, had all the faults of pre-Galilean science. Grant Aristotle his assumption—that a vacuum in nature is impossible, for instance—and what follows is logical. But Galileo had shown that Aristotle's physical presuppositions were wrong by means of observation. Torricelli was denying that it was legitimate to use a similar reference to observation as a criticism of Galilean ballistics, and re-opening the breach between scientific speculation and physical reality which it had been Galileo's effort to close.[1]

This is not to say that either Galileo or Torricelli were excessively simple in their thoughts on motion, or that Galilean ballistics, because they were only true for conditions which never occur in nature, or because, far from being the result of experiment in a manner which has sometimes been said to be characteristic of modern science, experiment showed their inadequacy in detail, was wholly useless. On the contrary, Galileo's conceptions were essential if the science of motion was ever to progress beyond the theory of impetus. Galileo's great achievement had been to dismiss all existing classifications of motion and deal with it under its simplest definitions, as either uniform, rectilinear and eternal

[1] It is relevant to point out that the earlier portion of the treatise in which Torricelli speaks abstractly of 'projections' was written in Latin, but at p. 217 he commences to write in Italian and speak of guns and artillery. Some idea of the practical nature of this part of the work may be given by illustrations. 'Let us suppose that the longest shot, that is to say that made at the elevation of the sixth point of the quadrant, made by a culverin is for example, 4000 geometer's paces. I wish with the same gun to make a shot of 2360 paces . . . [p. 217]. Let a piece be aimed along the line AC of which the elevation is the angle BAC, whatever that may be. The angle is measured with the quadrant and found to be, for example, 30 degrees, then the gun is fired, the ball strikes at the point B, and the distance AB is measured as, for example, 2400 paces [p. 218]. It happens many times that one has to shoot against a plane perpendicular to the horizon, such as the wall of a city or a tower or something else. Then we can in this case also suggest a calculation for the height of the point where the shot will strike the wall . . . [p. 222]. Suppose that the mouth of a culverin is at A, BC the horizontal plane: and AB the height of the gun, 2 braccia . . .' [pp. 222–3]. Language like this might well deceive the Genoese gentlemen into thinking that the parabolic trajectory was more practical and less theoretical, than Torricelli afterwards asserted.

or uniformly accelerated, rectilinear and eternal. If the motion of a body was compounded of these simple motions then (as Cavalieri and Mersenne saw also) it would move in a parabola; this was the simplest possible way in which a projectile could move in space empty of all but 'forces'. It was essential to know this before the genuine study of the movement of projectiles could begin; it is still taught in the introduction to text-books on ballistics. Science could only progress by moving from the simple unreal to the highly complex real and Galileo's virtue was to separate those stages. The mistake which led Torricelli into his defence of Galilean ballistics was made by those who underestimated the degree of unreality in all Galilean physical hypotheses, who believed that the problem was already solved with sufficient approximation. Here Galileo himself must be charged with confusion, for it is not at all clear that he was convinced that his great discovery was useless for practical purposes; as for Torricelli's assertion that neither he nor Galileo ever believed that the science of motion had anything to do with practical affairs, it is contradicted by Galileo's letters and his own plain use of terms.

The history of ballistics thus far affords an example of the unfolding of solutions to a group of related mathematico-physical problems whose intractability appeared almost axiomatic until 1632, which in ten years was complete and, indeed, within the original limitations of the problems, the solutions have been only restated in form during the succeeding three centuries. There are two salient features of this stage; first, the fundamental difficulty lay in the formulation of new concepts according to which natural phenomena could be classified and interpreted; second, the parabolic trajectory of Cavalieri, Galileo and Torricelli and their successors was precise and invariable, in it the mathematical elements were more important than the physical.

The progress from the theory of impetus had demanded a great philosophic effort, the tension of creative thought, rather than a straining of mathematical knowledge, for the parabola was one of the most familiar of curves and ballistical theory was constructed on its defining property that the amplitude varies as the square root of the diameter. This was what Galileo meant when he wrote that it was hard to arrive at the idea that a projectile must move in a parabola yet easy to demonstrate that it must. Moreover the parabolic trajectory was immutable; only its scale, not its shape, was affected by alterations in the angle of projection. In its very conception it was free from the embarrassment of physical complication and in theory one observation alone was necessary to ascertain all possible ranges. The parabolic theory of projectiles appeared to provide a clear road through all the difficulties which had surrounded the philosopher in his thoughts on motion since the time of Aristotle, and more practical men looked to it for the same assistance. It was in vain that Torricelli declared that the question 'Is it true?' was irrelevant; the geometrical and hypothetical truth he had assigned to it was not enough. Others at once began to apply to Galilean ballistics the test of experiment which Galileo had applied to Aristotle's science of motion.

CHAPTER V

MATHEMATICAL BALLISTICS II

The Resistance of the Air

It is inevitable in the progress of a science that as the mass of experimental knowledge grows its synthesis within the existing primitive scheme of ideas will present increasingly greater difficulties; discordant or conflicting observations will become more numerous, until at last order can only be restored through the evolution of a new concept making possible more comprehensive generalisations about nature. Such generalisations which build up the framework of thought are clearly incapable of exact laboratory proof. A revolution of this sort in dynamics was effected by Galileo in the early seventeenth century, with the support of other less influential but equally original philosophers like Isaac Beeckman in Holland and Kepler in Germany, at the cost of a change of emphasis. Air resistance, the point on which medieval dynamics turned, was of little moment to him. His purpose was to establish not the exact, but the fundamental theory of motion. To the more limited branch of science called ballistics belong the problems which are encountered in the world of experience where projectiles move through a medium of varying density, over a surface which is not plane but spheroidal, and so forth.

In this early period the distinction between the fundamental theory and the detailed examination of phenomena, partially obscured by the crudity of observation, was not so apparent as it is to us. Philosophers were expected to produce results which could be applied with the rigidity of an engineering formula. The old idea of the perfection of nature survived, so that men could still readily believe that the heavenly projectile moved in a perfect circle, the earthly one in a perfect parabola. Consequently hypotheses based upon simplified ideas of nature were thought to be accurate enough for practical use until experience pointed the contrary. Galileo and Torricelli, setting out to correct the dynamical doctrines of philosophy by the experimental and mathematical methods of the study, gave an impression in their works, for which there was reason, if not excuse, in the contemporary utilitarianism, that the ancient problems of the gunner and the philosopher had been solved together. But to neglect the resistance of the air in theoretical science did not mean that it was also negligible in applied science. Galileo had reached his greatest success by taking his particles and forces into free space; the ballistician could only succeed by remembering that for him they operate on the earth's surface.

Galilean dynamics denied two important philosophic assumptions: the first, that air resistance has important effects on the motion of heavy bodies, had been accepted by Aristotle and even widened by the impetus school; the second, that the universe is a plenum, had been almost unchallenged since late classical times and was revived in the philosophy of Descartes.[1] The price paid for the establishment of the fundamental laws of dynamics was the abandonment of the common-sense survey of nature which was at the back of these assumptions. To geometrise gunnery in accordance with the parabolic theory and on the hypothesis that the movements of particles in empty space could serve as a guide to the motions of solid bodies on earth, it had to be proved that, for ordinary projectiles at least, Galileo's law of acceleration was not vitiated by air resistance, that the horizontal velocity of a body remained constant, however swiftly it moved, and that the medium did not in any way interfere with the compounding of vertical and horizontal motion.

Cavalieri simply thought that the disturbing effects of air resistance were not great, 'che per esser tenuissima e fluidissima, per qualche notabile spatio, piu esser, che gli permetta la sudetta uniformita'.[2] Galileo, however, made a more extensive study of air resistance in his discussion of the rate of acceleration of falling bodies, where, in his demonstration that all, whether light or heavy, fall at the same increasing speed and strike the gound at the same instant *in a vacuum*, he had clearly distinguished the effect of air resistance, in any practical experiment causing the

[1] See below, p. 109. [2] *Lo Specchio Ustorio*, p. 157; cf. also pp. 161, 164, 169.

denser body to outstrip the less, a piece of lead to fall more quickly than a bunch of feathers. But, he said, let not the Aristotelians try to hide the folly of their contention that the velocity of descent is proportional to the weight of the falling body behind the fact that a cannon-ball reaches the ground a mere inch before a musket ball when they are dropped together from the same height. Galileo knew that air is a fluid having weight, and he supposed that just as cork is more buoyant than iron in water, it will be so in air too (even though air is some hundreds of times less dense than water) and for that reason it will fall at a slower rate. His observations satisfied him that in practice all materials whose density is very great compared with air—wood, stone, metal—fall at approximately equal speeds; and equal pendulums of these materials vibrate in almost equal times. At ordinary velocities, therefore, including those of projectiles propelled mechanically, air resistance might safely be neglected, and, in order to adopt a scientific procedure of abstraction, the theorems of motion should first be demonstrated without considering resistance, and then later modified as necessary in accordance with experience.[1]

Though this discussion, in which he shows that the errors in his strict theory of falling bodies are slight and attributable to air resistance, seems at first sight to indicate that Galileo was inclined to minimise the important alterations in the simple parabolic theory of projectiles caused by the resistance of the air, he was well aware of another theoretical possibility. Suppose a body to be accelerated indefinitely through a resisting medium by a constant force; then if the increasing resistance at last becomes equal to the force, the body reaches a terminal velocity; it can move no faster. Galileo believed that the 'supernatural' velocity given to a projectile by the force of gunpowder was already greater than its terminal velocity, and that consequently it must be retarded by the air from the very moment of firing, so that its path through the air would not be a true parabola, but a curve more steeply curved at the end than at the beginning. Yet he thought that this modification was not so important that his theory of projectiles was of no practical use, especially as this would be most

[1] *Opere*, vol. VIII, pp. 106–9, 118–21, 129, 136, 276–7.

properly applied to shooting with mortars, where the velocity of projection is low, and the perturbing effect of air resistance small.[1] In a letter of January 1639 he made the practical application quite plain, remarking that he had proved that a gun shoots its greatest range when elevated to forty-five degrees.[2]

The theory of the parabolic trajectory was not asserted to be an entirely accurate representation of the phenomena nor a merely hypothetical picture, but a very close approximation based upon the most satisfactory laws of mechanics with indisputable mathematical logic and capable, therefore, of application in practice. There was no inherent improbability in this. As it was known that a piece of wood in falling could attain an apparently enormous velocity without revealing any retardation by the air, it was reasonable to suppose that such a small heavy body as a musket ball could suffer none. Even in the sixteenth century, from looking at the flight of fire bombs, it had been guessed that they followed a symmetrical curve.[3] Finally, as the resistance operated on both the components of motion alike, was it not likely that though the parabola might be smaller, it would not be vastly changed? It must be remembered that, as the primary difficulty overcome by the author of the *Discourses* had been philosophic, so when the task was transferred to the mind of the reader the problem of acceptance was philosophic rather than practical. To the seventeenth-century mind the reception of Galileo's dynamical principles involved not only the sacrifice of all traditional learning but a tremendous effort of comprehension, of which the measure is the length and diversity of Galileo's exposition in the *Dialogues* and *Discourses*: who, then, having made the crossing between medieval and modern dynamics which almost made the mind reel, was to fear to cross another little chasm in the path of the master? The simple conclusiveness of Galileo's argument seemed to offer the clue to the understanding of all forms of motion. Only experience could show that the application of natural laws may mean the virtual abandonment of them and an attack on very different problems. In the theory of projectiles the difficulty was

[1] *Opere*, vol. VIII, pp. 137-8, 278-9. [2] *Ibid.* vol. XVIII, pp. 12-13.

[3] See for instance the drawings made by Leonardo in the Codex Atlanticus and the frontispiece to Tartaglia's *La Nova Scientia*.

particularly great, for whereas the laws of acceleration and inertia are straightforward and axiomatic, to calculate the allowances to be made for the resistance of the medium requires complex mathematics and exact experimental method. The failure of Galileo and Torricelli to verify the theory of projectiles by experiment or even to compare it with the results obtained by the students of gunnery, obscured the deficiencies of a theory which was from the *prima facie* evidence extremely plausible.

A short-lived criticism of it came from the northern countries where lay the focus of scientific activity with the drawing to a close of Galileo's active life as a philosopher. Neither Mersenne, Fermat, Roberval or Descartes believed that there could be entire truth in theorems dependent on such an extreme simplification of the normal world of experience.[1] The French philosophers grouped around Mersenne in Paris were pervaded by the influence of the Rector of the school of Dordrecht, Isaac Beeckman, a mentor to both Mersenne and Descartes. He was among the first to stress the importance of air resistance in the theory of dynamics, to refer to terminal velocity ('*punctum aequalitatis*') and attempt to demonstrate it experimentally. Even in making experiments with pendulums, he said, it was necessary to remember that they were swinging in air, not a vacuum. He explained that small bodies are more resisted than large because weight increases as the third power of the diameter, area only as the second.[2] Mersenne, undertaking the task of interpreting the work of Galileo to Frenchmen, was also in close touch with what was, perhaps, the next most original mind of the age, and his book on ballistics was much more than a mere exposition of the Galilean theory of projectiles.

In fact *Ballistica et Acontismologia* (1644) not only coined the name for a new branch of mechanical science; it originated the connected study of air resistance. Neither a very original philosopher, nor a constructor of physical theories, Mersenne possessed a highly developed curiosity and a talent for experiment. He wrote more accounts of things done with his own hands and seen with

[1] Cf. e.g. the correspondence quoted in *Opere*, vol. XVII, pp. 369, 387, 394, etc.

[2] Waard, Cornelis de (ed.), *Correspondance du P. M. Mersenne* (Paris, 1932), vol. II, pp. 122, 219, 233, 236, 237–8. Cf. also p. 340. There are four letters from Beeckman in this volume.

his own eyes than anyone before the great age of the scientific societies. By observing that the impact of a musket ball against a wall is heard by one stationed on the far side at the same instant as the report of the gun, he showed that the velocity of the ball must be of the same order as the velocity of sound, which he believed to be about 1920 feet per second. The speed of larger missiles he placed at 600–900 feet per second.[1] As an estimate of the possible effect of the resistance of the air he worked out that the musket ball, having a velocity of 600 feet per second must displace 14,400 times its own volume of air in each second of its flight.[2] Only if a heavy body in falling continued to accelerate ad infinitum in accordance with Galileo's law would a projectile move in a perfect parabola, but Beeckman had proved to him that this was impossible. Like Torricelli he tested the theory of projectiles in experiments on jets of water and found the effect of air resistance appreciable, since the curve of the jet was not perfectly parabolical, the falling water tending more to the vertical. He also satisfied himself in trials with a crossbow that the time of ascent of a projected body is always less than the time of descent, the difference increasing with the velocity of projection.[3]

Mersenne did attempt to discover one or two mathematical relations, but he was finally forced to admit that he could not plumb the problem. He said, correctly, that the air resistance increased with the velocity of the projectile, and inversely with its radius and density, explaining thus the relatively longer ranges of large cannon, and the slower fall of cork compared with wax, or pith compared with cork.[4] On the other hand he thought that there was some simple relation between the height and amplitude of the trajectory in a resisting medium as in a vacuum. He also printed one of the 'practical' range tables (actually a slightly involved arithmetical progression) and pointed out that it was very different from that derived from theoretical considerations by Galileo, but could not account for the disagreement.

[1] *Reflectiones physico-mathematica* in *Universae Geometriae Mixtaeque Mathematicae Synopsis* (Paris, 1644), pp. 127–8.

[2] *Cogitata physico-mathematica* (Paris, 1644); *Hydraulica*, Preface, sig. e iii.

[3] *Hydraulica*, pp. 109, 116, 135; *Ballistica*, pp. 24, 31, 34, 83. In this Mersenne owed much to Descartes, e.g. *Œuvres* (Adam-Tannery) vol. III, p. 657; vol. IV, p. 687.

[4] *Ballistica*, pp. 132, 137 *et seq.*

Although the deduction was clear from Mersenne's work that the simple parabolic theory of projectiles must be modified to make allowance for the resistance of the air, he had scarcely advanced a step towards the solution of the problem thus raised. His long and diffuse writings, in which the reader could easily lose sight of the main points amid frequent repetitions, mistakes and digressions, were not likely to be highly esteemed when a straightforward geometrical construction was sought: the seventeenth century asked more of a philosopher than the mere statement of experimental results, it expected that they should be fitted into some general hypothesis or system to include all related phenomena.[1] Mersenne was usually described as an intelligencer and praised for his part in stimulating scientific discussion, gathering mathematicians and philosophers together. His own experimenta researches were overlooked. Most thinkers—even Descartes—became reconciled to escaping the complexities of nature by writing of Galilean dynamics as though they were observationally exact.

Scientific ballistics remained purely theoretical. Galileo's very great renown and the logic of his method was the first reason fo this; another was the triumph of Cartesian philosophy, in its return to the belief that the great truths of the universe could be better reached by intuitive reasoning than by laborious mathematical analysis of all its many parts. But the most powerfu reason for the lack of progress in ballistics was the absence of a suitable mathematical technique. Any geometer could now 'prove Galileo's theorems; no one knew how they should be re-interprete in a theory of resistance. No one had even given a mathematica expression for resisted motion.

Nevertheless new knowledge which prepared the way becam available. Descartes contributed a hypothetical explanation of th movement of a body through a resisting medium. He describe all bodies as consisting of innumerable infinitely divisible particles In a perfect medium—the aether—these all move at a grea velocity, so that an immersed body is subjected to an equal pressure on all sides, and in whatever direction it moves, the impac

[1] Compare the early reception of Newton's papers on light; Huygens' words on Boyle 'Il paroit assez étrange qu'il n'ait rien basti sur tant d'experiences dont ses livres son pleins. . .' (To Leibniz, 4 Feb. 1692 [*Œuvres*, vol. x, p. 239]).

of the particles remains the same because their movement is so much more rapid than that of any solid body. Thus a perfect medium offers no resistance to motion. Air, water and all physical fluids contain much grosser corpuscles of a more sluggish movement which, lying in the path of a moving body, cause it to lose its velocity by impact against them. The moving body tends to compress together the corpuscles in the direction of its motion and to leave a rarefied space behind; to prevent this unnatural condition—for according to Descartes the same quantity of matter must always occupy the same space and even God could not create a space empty of all matter—the particles must flow round the body from front to rear causing a circular eddy about it. Force is required to move these heavy particles which distinguish ordinary fluids, and this can only come from the moving body itself. The more rapid the motion of the body, the more quickly its motion will be dissipated in stirring up the corpuscles.[1]

By this theory—which is, dynamically, a transposition into three dimensions of the familiar events seen when a stick is moved through water covered with bits of dust and straw—all resistance of the medium was reduced to a single physical phenomenon, the impact of groups of particles. It was a theory having all the merits of the corpuscular philosophy, though Descartes himself always denied that philosophy in its more extended and especially chemical sense as used by Gassendi and Boyle. It had the advantage of using the well-known law of inertia, now widely accepted, to explain several phenomena of resistance. The inertia of the corpuscles of fluids caused the loss of motion by a moving body; the inertia of the moving corpuscles was the reason for the swirls and currents in the fluid persisting after the passage of the moving body. Further, it explained why the resistance was in some way proportional to the velocity. In fact it followed that the resistance must be as the second power of the velocity of the projectile. For if at a velocity V it penetrates a distance S through the medium

[1] *Principia Philosophiae* (1644) in *Œuvres* (Adam-Tannery, Paris, 1905) vol. VIII, pp. 50, 53, 70 *et seq.*; *Le Monde* (1664) in *ibid.* vol. XI, p. 19; letter to Sir William Cavendish, Nov. 1646 in *ibid.* vol. IV, p. 559. Descartes' ideas were thus in complete opposition to the note made by Leonardo: 'The air becomes condensed before bodies that penetrate it swiftly, acquiring more or less density as the speed is more violent or less' (MS. E, ov., 73r.).

in one second, striking N corpuscles, at a greater velocity nV it will pass over a space nS striking nN particles and each impact will be n times more violent. Therefore the loss of velocity and the resistance will be multiplied by n^2.

Descartes also proposed in a letter to Mersenne a possible mathematical representation of the fall of a heavy body through a resisting medium which he seems never to have developed further, though it was put in print by Mersenne. Suppose a body to fall so that its velocity at the end of successive seconds is g, $2g$, $3g$, . . . etc. but that also during each second the resistance reduces the velocity by a factor R; then after the first second the whole velocity will be $\frac{g}{R}$, after the second $\frac{g}{R}\left(1+\frac{1}{R}\right)$, after the third $\frac{g}{R}\left(1+\frac{1}{R}+\frac{1}{R^2}\right)$, etc. Since the term within the bracket is a geometric progression with a definite sum for however many seconds the fall may last, Descartes claimed, there could be no terminal velocity. The falling body continued to accelerate though the rate to acceleration became very small. Only after an *infinite* fall would a maximum velocity $\frac{g}{R-1}$ be attained, therefore in practice the terminal velocity could never be found.[1]

In contrast to the speculative character of Cartesian philosophy were a number of researches in experimental physics which also promoted the study of air resistance. In the middle of the century the invention of the air-pump by Otto von Guericke, who demonstrated the pressure of the air most strikingly with his 'Madgeburg spheres', and its adaptation into an efficient laboratory instrument by Hooke for Robert Boyle were great steps forward in aerostatics. A description of numerous experiments with it was published by Boyle in 1660. The density of air and its resistance to compression or dilatation (which Boyle called its 'spring') were easily proved with the aid of the pump, and a great deal of new evidence was brought forward to show that the air was not an inert substance but matter with important chemical, physical and mechanical properties. Boyle even devised the famous 'guinea and feather' experiment to illustrate in a limited space the fact that

[1] Mersenne, *Correspondance*, vol. II, pp. 342 *et seq.*; *Harmonie Universelle* pp. 206–7.

two bodies of very different densities fall at the same speed only in a vacuum.[1] Robert Hooke further suggested in 1663 that the effect of air resistance could be measured experimentally in the glass vessel of a reversible air-pump by counting the number of oscillations made by a pendulum in air at reduced, normal, and increased pressure.[2]

At about the same time too many attempts were made to verify Galileo's law of acceleration by direct experiment. He himself had proposed a method of illustrating rather than proving it by measuring the time taken by a sphere to roll down an inclined plane, which Mersenne had found to be very uncertain. Other measurements had been made by G-B. Riccioli in dropping balls of wax, wood, iron, etc. from the top of a lofty tower in Bologna.[3] The Royal Society began experiments on the velocities of descent of various substances in air and water, of which the conduct was entrusted to Lord Brouncker, in October 1661. Hooke devised several machines for measuring the velocity of a falling body directly or by measuring the impact, and in March 1663 read a long memorandum of suggestions for ascertaining the velocity and force of impact of projectiles. Again in the following year the Royal Society was 'much engaged in the theory and experiments of the descent of bodies and their weight in several mediums'.[4]

The great Dutch physicist Christiaan Huygens was drawn into these investigations by his friend Sir Robert Moray, the first president of the new 'College for the promoting of Physico Mathematicall Experimental Learning' before it became the Royal Society. In 1646, at the early age of seventeen, Huygens had worked out an independent mathematical proof of Galileo's law of acceleration, and in 1659 he had satisfied himself that the terminal velocities of two similar bodies are equal if their radii are inversely proportional to their densities.[5] At this time he did not think that there was any real need for a modification of the simple law of

[1] *Works* (ed. Thomas Birch, 1772), vol. I, p. 61; vol. II, p. 256.

[2] Robert Gunther, *Early Science in Oxford*, vol. VI, p. 115.

[3] *Almagestum Novum* (Bologna, 1651) pp. 89 *et seq.*

[4] Birch, *History of the Royal Society* (London, 1756–7), vol. I, pp. 46, 49 *et seq.*, 195, 205; Gunther, vol. VI, p. 200.

[5] *Œuvres Complètes*, vol. I, p. 24; vol. XI, p. 72; vol. XVI, p. 254, 256, etc.

acceleration, although it could be shown to be theoretically imperfect for a resisting medium. This opinion he expounded to Moray in 1667.[1]

Huygens had already obtained an even more important result—an exact value for the gravitational constant (g) which must appear in every ballistical calculation. His evaluation of g was completely independent of any experimental measurement of the acceleration of falling bodies over short distances (where minute errors in time are important) or over long distances (when air resistance becomes a significant factor). He discovered that the time of oscillation of a simple pendulum is given by $t=\pi\sqrt{\frac{l}{g}}$ where l is the length of the pendulum beating seconds. From his experiments he found that $g=31{\cdot}25$ Rhenish feet per second, equal to 981 cm/sec. We may judge that Huygens was extremely proud of this research since he only allowed it to become public by slow stages; the exact value of g was sent to Moray in 1664; the formula was communicated officially to the Royal Society as an anagram in 1669, and finally published in 1673.[2] The importance of knowing the precise rate of acceleration under gravity in any attempt to determine the effect of air resistance on a moving body is obvious enough, but experiments with falling spheres to check a scientific theory of resistance were not made before those of 1710 by Newton and Francis Hawksbee.[3]

The yielding of the secret of the gravitational constant, in itself a notable scientific achievement, in ballistics was for the present only a contribution towards the study of the purely theoretical parabolic trajectory. More significant was the fact that now the mind of one man held all the knowledge necessary for the solution of the simpler problems of ballistics, and was conditioned towards interpreting observed facts in the one way which would bring that solution within his grasp. Huygens knew how a projectile would move in empty space: he knew how to represent changes in velocity, time and space geometrically; he knew that the inertial and the gravitational velocities were modified at every

[1] *Œuvres Complètes*, vol. III, pp. 317 *et seq.*

[2] *Ibid.* vol. v, p. 84; vol. VI, p. 490; vol. XVI, p. 280, 410; vol. XVII, p. 100, 245-7, 278-84; vol. XVIII, pp. 354 *et seq.*

[3] *Philosophical Transactions*, no. 328 (Oct.-Dec. 1710), pp. 196 *et seq.*

moment along the path of the projectile by the resistance of the air. Cartesian geometry supplied a method of describing the resultant curve, however irregular, by plotting the position of the projectile at any moment by reference to horizontal and vertical coordinates. The difficulty of calculating the coordinates remained. If they could be defined mathematically and placed in some certain relation to each other the problem was solved.

For Huygens, as earlier for Beeckman, who had remarked that the terminal velocity of a body would be the same as that of an upward current of air which would just prevent it from falling, resistance was a 'pressure' like that of gravity. He did not primarily consider it as a product of the corpuscular structure of matter in the way described by Descartes.[1] In their relative movements, whether the body or the medium is moving while the other remains at rest, the surface of the body is pressed upon by the medium. Since the pressure of a moving current of air is in some way proportional to its velocity, conversely the resistance of a motionless medium to a projectile must be proportional to the velocity of its penetration. This variation from moment to moment of the relative speeds and resistances was at once a complication ignored by his predecessors and a step towards the mathematical analysis of the problem; so long as the weight and size of the projectile and the resisting powers of the medium remained the same the retardation at one speed could be calculated directly from the retardation at another.

Huygens' notes on motion written in 1668-9 show the progress of his thought from the simple to the complex, until finally he had to admit defeat. He first calculated the height which a body would reach when projected vertically upwards at its terminal velocity in a medium resisting in direct proportion to the velocity of the projectile. This hypothesis had the advantage of giving the retardation in the first instant as $2g$, the resistance and the force of gravity being equal. His method was to compare the movement of the resisted body with that of one assumed not to be resisted, the decrement of velocity of the former being twice that of the latter at the start of the upward motion and equal to it

[1] *Œuvres*, vol. X, p. 19; vol. XIX, p. 79.

(i.e. g) at the end. He showed by analysing the motion geometrically, placing his time axis vertically and taking retardation including the force of gravity as his horizontal axis, that the decrease in velocity of the resisted body followed a logarithmic curve, and from this was able to calculate the area representing the height it attained.[1] Under these conditions the maximum height of the body in the medium is $0{\cdot}308\frac{V_t^2}{g}$.[2] Next in the examination of vertical descent he found that in any time t the resisted body fell the distance $\frac{gt^2}{e}$ or $0{\cdot}367\,gt^2$. Finally he discovered a construction for tracing the trajectory of a projectile when fired at 45° inclination to the horizontal, assuming that the resistance of the air varied directly as the velocity and that the velocity of projection was $\sqrt{2}V_t$.[3]

His work of the following year (1669) reveals that Huygens was not long content with this simple theory that the resistance was directly proportional to the velocity, and he set to work to establish by a series of experiments a different hypothesis which he had come to think far more probable, namely that the resistance was proportional to the square of the velocity. It is likely that he was led to this view by *a priori* considerations and sought to verify it by tests; the records of his experiments suggest very clearly the confirmation of expected results.

In his first experiments he arranged for a jet of water to impinge on one arm of a balance, the other being held down by weights; it was discovered that the force of the water on the balance was roughly proportional to the height of the head of water causing the jet, and since the velocity of the issuing water was (by Torricelli's law) itself proportional to the square root of the height, the force of the water varied as the square of its velocity.[4] The resistance of still water was estimated by an even more

[1] *Œuvres*, vol. XIX, pp. 102 *et seq.* Huygens' geometry is equivalent to integration of the equation of motion $\frac{dV}{dt} = -g - kV$, giving $V = \frac{g}{k}(2e^{-kt} - 1)$, g/k being the terminal velocity. At the vertex $V = 0$, whence $2e^{-kt} = 1$ and $e^{kt} = 2$. Therefore $t = \frac{\log_{10} 2}{\log_{10} e}$ (*ibid.* pp. 83–4).

[2] I.e. $\frac{V_t^2}{g}\left(1 - \frac{\log_{10} 2}{\log_{10} e}\right)$ which is more accurately $0{\cdot}307\frac{V_t^2}{g}$.

[3] *Œuvres* pp. 115–18.

[4] *Ibid.* pp. 120–1, 123–4 (3 April 1669).

ingenious apparatus: a long trough was filled with water and a block of wood attached to a cord floated in it. The cord passed over a pulley at the end of the trough to a weight which, as it fell, drew the block through the water. After allowing a little distance so that the weight could attain its full speed, the number of seconds required to draw the 'boat' the whole length of the trough were counted, using different weights to provide the motive power. With weights of $\frac{1}{2}$, 1, and 2 ounces the times were 11–11$\frac{1}{2}$, 7$\frac{1}{2}$, and 5$\frac{1}{2}$ seconds. Thus when the weight was doubled the velocity was increased by $\sqrt{2}$ approximately, or in other words the resistance of the water overcome by the weight increased as the square of the velocity of the 'boat'.[1] The same idea was tried in the Seine at Paris, where Huygens then was, by finding out what weights were required to keep a float stationary against currents in the river of varying speeds.

The last of Huygens' experiments was the most elaborate and would have been most to his purpose had it succeeded fully. Two light screens on little carriages were arranged so that by means of cords and pulleys of different sizes one could be pulled along twice as fast as the other. Each screen was free to swivel over backwards when the pressure on its face was sufficient to overcome the retaining force of a small weight. The weight on the faster-moving screen was four times as heavy as that on the slower and Huygens' intention was that the screens should be pulled along more and more quickly by the pulleys until the air resistance would cause both screens to topple over at the same instant. Trials were not very successful owing to maladjustments, but were good enough to satisfy Huygens that the square law was accurate.[2]

So strong was his confidence in a single law of resistance that he felt no hesitation in extending it from the very low velocities of which he had some experience to the much higher velocities of ballistics of which he had none at all, and from the fluid to the gaseous state.[3] He was doing more than assign a value to the index in the expression $R=KV^n$; he was illustrating and

[1] *Œuvres*, vol. XIX, pp. 122, 124–5. [2] *Ibid.* vol. XIX, pp. 138 *et seq.*

[3] Although his father Constantine spent a good part of his life in camp and carried out some experiments on artillery for Mersenne, there seems to be no record that Christiaan had any acquaintance with military life and affairs.

confirming a deduction from the essential simplicity of nature, that the square law was as appropriate for resistance as for acceleration, or (as Newton was later to prove) for gravitational attraction. This particular law of resistance should therefore rather be associated with the name of Huygens than that of Newton. He thought that if experimental results were never exact, this was more probably due to inadequacies in the method and the apparatus than to such a complexity of nature as n itself being a variable.[1]

After adopting a new theory of resistance Huygens naturally revised his ballistical calculations, again using as the chief tool of his analysis the logarithmic curve, by means of which he could evaluate areas and so integrate the equations of motion. As before, he assumed the velocity of projection to be the terminal velocity for the sake of simplicity so that initially $R=2g$.[2] He calculated that the time of ascent in a medium resisting as the square of the velocity (V_t) was $a^2=(1-\frac{1}{3}+\frac{1}{5}-\frac{1}{7}+\ldots)$ where a^2 was the area representing the equivalent time of ascent in vacuo. Only in 1674 did he learn from Leibniz that this series expressed the ratio of the area of the circle to that of a circumscribed square, so that the time of ascent in a resisting medium of the type he was considering could be more conveniently written $\frac{\pi V_t}{4g}$. With this knowledge he was able to reduce the problem of finding the height of ascent to one of the quadrature of hyperbolic areas—for which several methods were already known, especially those of Mercator and Wallis using infinite series—and found the height to which the projectile could ascend against resistance to be $\frac{0{\cdot}693148}{2g} V_t^2$ approximately.[3]

These were the first-fruits of Huygens' endeavour to formulate a more rigorously exact science of ballistics. To attempt to calculate a trajectory in accordance with his later hypothesis of resistance proved beyond his powers, as it was later to prove too difficult

[1] Cf. the remarks of Huygens' editors (*Œuvres*, vol. XIX, p. 85).

[2] $V_t^2=\frac{g}{K}$. For ascent $\frac{dv}{dt}=-g-Kv^2$; for descent $\frac{dv}{dt}=g-Kv^2$. Initially $\frac{dv}{dt}=-2g$ since $K_v^2=g$ by definition.

[3] *Œuvres*, vol. XIX, pp. 144 *et seq.*

for Newton's method of fluxions. In addition Huygens had realised that if the medium resisted as the square of the velocity, then the velocity of an oblique projection could not be analysed into horizontal and vertical components: the vectorial sum of the velocities would be correct but not the sum of the resistances corresponding to them.[1] At this point he lost interest in a problem whose intrinsic interest appeared small compared with the immense difficulties in the way of a solution. All these researches, which give Huygens a definite priority in the study of resisted motion, remained unpublished for more than twenty years 'ayant négligé de les achever parce que cette speculation ne m'a pas semblé assez utile ni de conséquence à proportion de la difficulté qui s'y rencontre'.[2] Huygens was aware that he had exploited his mathematical technique to the full and its limitations had become apparent. 'Il est extremêment difficile si non du tout impossible de resoudre ce problème,' he wrote in 1690.[3]

In his correspondence with Leibniz on the differential calculus it was only slowly that Huygens could be persuaded that there were methods of analysis superior to his own, and indeed his results in 1669 are a wonderful achievement. He had discovered much that only became apparent to Newton through the use of fluxions and before the publication of the *Principia* no one else had done nearly so much as he. At this time only the first steps towards the differential and integral calculus were being taken by Barrow, Gregory, Newton and Slusius; it was still far from being the precise system into which it developed in the hands of Leibniz and his pupils. No one who has looked over the pages in which Huygens developed his dynamical theorems printed in the *Œuvres Complètes* can avoid admiring the acumen of Huygens' reasoning or the persistency which carried him through a long and complicated process, but a glance at the footnotes shows the convenience of the Leibnizian notation from which Huygens' hard-won relations appear almost at once. As Huygens knew, the progress of ballistics beyond the level which he had reached—which no one else was to equal for twenty years—depended on the invention of mathematicians.

[1] *Discours de la Pesanteur* (Leyden, 1690), p. 175. See below p. 147.
[2] *Ibid.* p. 169. [3] *Ibid.* p. 175.

Always provided that mathematicians wished to turn their attention to this type of problem. The question must be discussed more generally later; what was the purpose of Huygens himself in making these experiments and calculations? In several of his sketches appears a rough outline of a gun barrel at the point of projection; one cannot declare outright that Huygens' interest was purely abstract and that consideration of the practical form of projectile never entered his head. On the other hand there is no evidence that Huygens attempted to make his resistance experiments more practical or that he wished to do more than establish a theoretical law of physics. In his experiments on the resistance of water to the motion through it of a floating body he appears to have left aside altogether the possible interest of such researches for shipbuilding; for instance he never examined the effect of varying the shape of the floating block. Nor did he ever carry out any experiments with high velocities. The terms in which his calculations were formed were such as to render them quite useless for practical purposes, as he confessed. Nor did he attempt to translate them into practical language. Here of course he was just as much limited by his mathematical technique as by a possible lack of motive; he could not have done more had he wished. Huygens was not devoid of utilitarian motive or blind to the useful applications of science, for his whole work on the pendulum clock was directed towards the improvement of navigation, especially on long trading voyages. But when it is recognised that Huygens spent his early life in the atmosphere of the dispute between the Galileans and the anti-Galileans, that the question of resisted motion was one of the most keenly debated dynamical problems of the century, that Huygens himself loathed war and the state direction of science (which is probably the reason why he did not return to Paris after 1681), in Huygens' case at least there is no need to bring utilitarian motives to the fore in order to explain why he wrote on ballistics, above all when he so willingly gave up this work and buried it for twenty years.

Because Huygens' work was incomplete he made no public reference to it before 1690, though perhaps it is referred to obliquely in an offer he made in 1668 to Henry Oldenburg to communicate some of his discoveries to the Royal Society, and in *Horologium Oscillatorium* (1673),[1] yet in spite of the dominance of the parabolic theory which satisfied Blondel, Halley (for a time) and many others, there is scattered evidence that some men at least, though quite ignorant of Huygens' brilliant experimental work and mathematical researches (which indeed few were capable of rivalling), were not content with the parabolic theory and wished to examine the resistance of the air more thoroughly.

Apparently the Royal Society made no experiments on resisted motion between 1664 and 1668, but on the last day of that year Hooke is reported as saying that he 'conceived that the impediment given by the air or other fluids to moving bodies decreased in a continual proportion'; on which the President remarked that this ought to be shown by experiment.[2] In January 1671 the Society did in fact put to the test of experiment Galileo's assertion (fundamental to the parabolic theory) that if one heavy ball is allowed to fall, while a second is projected horizontally from the same point, they will reach the ground in the same instant. As in an earlier trial made by the Accademia del Cimento, the result was inconclusive.[3] Another experiment on the relation of resistance to area was demonstrated by Hooke in 1675, and in 1677 John Flamstead, the recently appointed Astronomer Royal, had some measurements made of the ranges of a steel cross bow at various angles of elevation which were later reported to the Royal Society by Collins and seemed to show a small but clear departure from the parabolic theory, ranges at elevations below 45° being always greater than those at the complementary angles above 45°.[4] Again at a meeting on 18 April 1678, Sir Jonas Moore assured the Fellows that in his experience of shooting mortar-bombs the greatest range was always at an angle below 45°. At twenty degrees of elevation the bombs would always fly farther than at

[1] *Œuvres*, vol. VI, pp. 276-7; vol. XVIII, p. 358.

[2] Gunther, *op. cit.* vol. VI, p. 347. Brouncker, Boyle, Moray and Hooke were members of a committee 'to consider of the improvement of artillery' (experiments being delegated to Hooke), March 1664/5 (*ibid.* p. 239). Perhaps Sir Robert Moray's *Experiments for improving the Art of Gunnery*, 'for the better determination of the true point-blank range, the optimum charge for any gun, and the farthest-shooting gun' should also be noted here (*Phil. Trans.* [June 1667] no. 26, p. 473).

[3] Gunther, *op. cit.* vol. VI, p. 373; Richard Waller, *Essays of Natural Experiments made in the Accademia del Cimento* (London, 1684), pp. 143 *et seq.*

[4] Gunther, *op. cit.* vol. VII, p. 434; Rigaud, *Correspondence of Scientific Men* (Oxford, 1841) vol. II, p. 172; Royal Society, *Classified Papers*, III (1), no. 15.

seventy, the reason for this being the density and resistance of the air to a body passing through it 'whereby that which was shot at 70° passing through a greater quantity of air received a greater impediment and hindrance from moving in a parabolical line than that which was shot at 20°'. The Surveyor-General went on to speak of the variable density and currents in the air as causes of errors in shooting, as well as the eccentric motion of the projectile. It was resolved that some experiments on air resistance should be made from the Monument on Fish Street Hill.[1]

In France Edmé Mariotte in his *Traité de Percussion* (1673) approached the study of the resistance of the air through the physical ideas of Descartes, making the last attempt to resolve the problem on the aerostatic lines suggested by Galileo. His results were not very illuminating, since they were virtually the same as those published by Huygens in 1659 relating to the resistance experienced by bodies of different sizes and densities, and had no general mathematical application. He did, however, affirm that he had found from experiments at the Paris Observatoire that a lead ball half an inch in diameter falls 14, 54, 117, 200, . . . feet in 1, 2, 3, 4, . . . seconds instead of the 15, 60, 135, . . . expected from Galileo's law of acceleration, and this led him to propose that the expression $s=\frac{1}{2}gt^2$ should be modified to $(\frac{1}{2}gt^2-xt^2)$ where x would be some factor depending upon the size and density of the resisted body, and the density of the medium.[2]

In the following year (1674) an English work on gunnery was published which soon became a standard authority, *The Genuine Use and Effects of the Gunne* by Robert Anderson. The author, described as a weaver by trade but skilled in the application of mathematics to gunnery and persistent in making artillery experiments at his own cost, was not unknown to more famous English mathematicians.[3] His attempts to use his knowledge of the art of calculation for the benefit of less learned gaugers and gunners—he seems to have been the first compiler of a gunner's manual to use logarithms—involved him in controversies with the rather

[1] Birch, *History of the Royal Society*, vol. III, p. 400. The next week the Society discussed the proximity of a trajectory to the parabola (*ibid.* p. 401).

[2] *Œuvres* (Leyden, 1717), pp. 74, 100, 103, 109–13. Here $x=\frac{1}{2}t+\frac{1}{2}$, and the theory breaks down after $t=19$.

[3] John Harris, *Lexicon Technicum* (London, 1704), *s.v.* 'Ordnance'.

quarrelsome Scottish mathematician James Gregory (1638-75), the inventor of the reflecting telescope. *The Genuine Use and Effects of the Gunne* is not a luminous book and some of the 'mathematical propositions' relating to internal ballistics are certainly foolish enough, but the theory of exterior ballistics, and the range tables calculated for various angles of elevation over level or sloping ground, are accurate—if the assumption is made that air resistance is negligible. This assumption Anderson claimed he had verified in his small-scale experiments, and it was this assumption that Gregory attacked. As John Collins, another 'amateur' of mathematics whose extensive correspondence was the innocent cause of the great dispute between Newton and Leibniz, remarked, in Anderson's book ranges of 8, 10 or 11 miles were quite usual, the projectile was made to follow an upright parabola, and ranges at elevations equally remote from 45° were said to be equal, in spite of experimental denials of these phenomena.[1]

The true reason for the notice taken of Anderson's book was the fact that James Gregory had himself published three years before an essay on ballistics, 'Tentamina quaedam Geometrica de Motu Penduli & Projectorum', in which he had renounced the original hypothesis of Galileo, followed by Anderson, and sought to calculate the trajectory of a projectile without neglecting air resistance.[2] Gregory supposed the air to be a uniformly resisting medium, so that a body moving through it would be subjected to a constant retardation, and based his calculations on the hypothesis that the motion of a projectile is compounded of a uniformly accelerated motion towards the centre of the earth caused by gravity (air resistance being negligible because the vertical velocity is small) and a uniformly retarded motion in the direction of propulsion.

Thus three forces operate on the projectile, its inertia carrying it along the line of the axis of the gun at an angle (α) of inclination to the horizon, air resistance causing a retardation (f) of the initial velocity of projection (v), and the force of gravity imparting to it an accelerated motion (g) vertically downwards towards the

[1] The correspondence between Collins and Gregory on ballistics and Anderson's book is printed in Turnbull, *op. cit.* pp. 282 *et seq.*

[2] Published anonymously as an appendix to a satirical pamphlet by 'Patrick Mathers' (William Saunders) called *The Great and New Art of Weighing Vanity* (Glasgow, 1672).

centre of the earth. Gregory's hypothesis may be explained rather more simply than it is in the 'Tentamina'. Suppose O to be the mouth of the gun, *OL* the direction of projection, and *K* the point where the ball first grazes the ground. During the time of flight, if its movement had been rectilinear, the projectile would have reached L in its retarded movement along *OL*, but because it falls

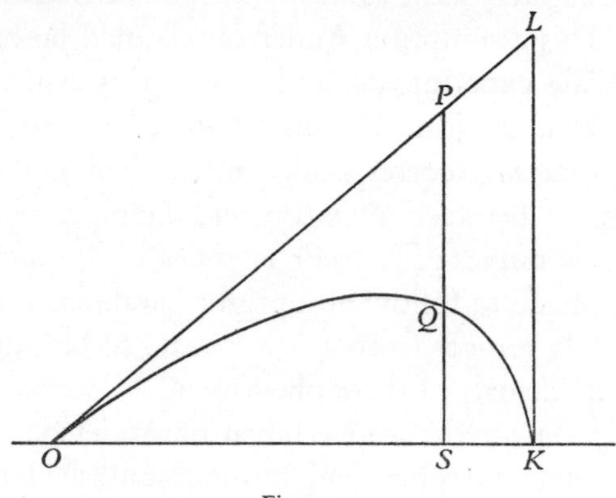

Figure 7

the perpendicular distance *LK* in that time, it strikes the ground at *K*, describing the curved path *OQK*. Since both the deceleration caused by air resistance along *OL*, and the acceleration due to gravity along *PS* may be calculated, the coordinates of any point such as Q on the curve may be found, and consequently the equation of the curve, which is a parabola whose axis is inclined to the horizontal.[1]

Obviously, if the simple parabolic hypothesis maintained by Anderson was correct in that air resistance might be neglected in calculation, Gregory was wrong; but his friend Collins tried to procure a general approval of his theory, and a condemnation of Anderson's, by sending copies of both books to such eminent mathematicians as John Wallis at Oxford and Isaac Newton at Cambridge.

Gregory did not easily arouse the support of other expert mathematicians, because his own views, which were as usual very difficult to understand, were scarcely more acceptable than those

[1] I owe the suggestion for the interpretation of Gregory's essay to Mr F. P. White of St John's College.

of Anderson. His hypothesis is indeed extraordinary, and very much inferior to that of Huygens, for not only does the retardation not vary with the velocity of the projectile, since it is proportional to the time of flight, but the resistance of the air is not considered to be opposed directly to the motion of the projectile. This, of course, would be of less importance if the trajectory were very flat. The proposition had rather a mathematical than a physical interest and was very soon forgotten; it is interesting, however, that Gregory's trajectory is very similar to that depicted by the impetus ballisticians.

In reply to Collins' request for a criticism of the rival theories of ballistics Wallis wrote a letter which summarises his own opinions at that time.[1] Anderson, he said, had followed the doctrine in Prop. 8 Cap. 10 of Wallis's own *Mechanics*, identical with that of Galileo and Torricelli;[2] while Gregory's theory might equally well be derived from the same work, since Wallis had himself remarked that the parabolic hypothesis could not be strictly accurate because the motion commonly supposed uniform is retarded by the resistance of the medium, for otherwise a bullet would strike with the same force afar off as close at hand, contrary to all experience.[3] And in practice, Wallis went on, cannoniers find the ranges of shots very different from the parabola, as in reason they ought to be, so that all such tables as Anderson's calculated on that basis are vain. As for Gregory's trajectory, though he had not examined it carefully, he doubted very much that it would prove to be an inclined parabola. This was not very flattering to Gregory, who retorted to Collins that all who thought so must be dunces and incompetent judges.

Newton, to whom also *The Genuine Use and Effects of the Gunne* was presented by Collins, replied from Cambridge on 20 June 1674, in much the same style as Wallis. This letter must be one of the earliest records of Newton's thoughts on mechanics. He said that Anderson's book was ingenious and might prove useful

[1] Dated 24 August 1647. Copy in Royal Society, *Letter Book*, vol. VII, p. 154; printed in Rigaud, *op. cit.* vol. II, pp. 587–9.

[2] *Mechanicorum sive Tractatus de Motu*, pars tertia, 1671; *Opera Mathematica* (Oxford, 1695), vol. I, p. 1001.

[3] The identical argument is used by Roberval in a letter to Torricelli in 1646 (*Opere*, vol. III, p. 354).

if all its principles were true, but he suspected the proposition that a bullet moves in a parabola. Newton's argument ran that if a bullet be shot horizontally from *A*, moving along the trajectory *AE* (*AI* being perpendicular to the horizontal), then it passes in equal times over the spaces *AF*, *AG*, *AH*, *AI*, which are in the ratio of the square numbers 1, 4, 9, 16; and drawing the horizontal lines through *F*, *G*, *H*, *I*, the bullet at these times will be located somewhere on the line, as at *B*, *C*, *D*, *E*. But that *FB*, *GC*, *HD*, *IE*, etc, are in arithmetical progression is not probable, because if

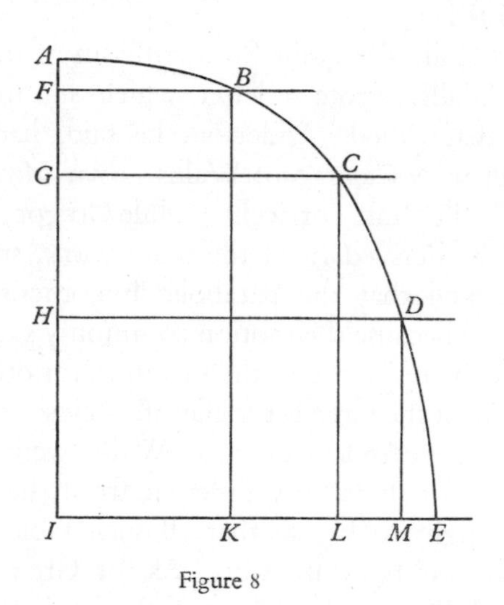

Figure 8

they were the velocity of the bullet must increase, since *AB*, *BC*, *CD*, *DE* described in equal times are successively larger; it is much more probable that owing to the resistance of the air the velocity of the bullet will decrease very considerably. Newton suggested therefore that a more close approximation to the descending branch of the trajectory might be obtained by dropping the perpendiculars *BK*, *CL*, *DM*, on *IE*, and making *IK*, *KL*, *LM* etc. in decreasing geometrical progression. He asked Collins, if he should speak of this idea to Anderson, to conceal his name as he had no mind to concern himself further in it. Busy as he was in defending his optical researches against short-sighted critics, he was not willing to allow another new idea to draw him into unwelcome controversy.

From this episode it may be seen that the most able British mathematicians were no longer content with the simple geometrical solution of the problem of exterior ballistics unfolded by Galileo, and that they were already aware of the necessity for some fresh mathematical treatment that should make due allowance for the progressive reduction of the initial velocity of the projectile caused by air resistance, if the question was ever to be raised above a very elementary level. Both Wallis and Newton seem to have had a sufficient sense of the mechanical difficulties involved to realise that Gregory's solution was not worth the trouble of detailed mathematical criticism, and though Gregory was still concerned for the fate of his 'Tentamina' up to the time of his death in October 1675, he failed to provoke a vigorous discussion; yet it is not impossible that this chance intervention, at the moment when Newton had temporarily put aside the investigations into gravity begun in 1666, may have stimulated his mechanical and mathematical researches into the problems of ballistics, on the widest scale, which were finally published in the *Principia*. The dispute between Anderson and Gregory had also more practical, if less permanent, results, for it induced Lord Brouncker, Sir Jonas Moore and other members of the Royal Society, with the assistance of the Master of the Ordnance, to carry out some experiments on the flight of projectiles upon Blackheath. Robert Hooke recorded in his journal three trials there during September 1674, when he found that the observations made on shooting with mortars did not greatly imperil the parabolic theory.[1]

By a coincidence, it was also in 1674 that Huygens added the final touches to his theory of ballistics, but the typical French work of this period was still of a more simple or practical quality. The celebrated and complete treatise by François Blondel, *L'Art de Jetter les Bombes*, was a direct response to Colbert's desire that the members of the Académie Royale des Sciences should devote themselves to researches of utilitarian value. Blondel (1617–86) was not himself a particularly notable mathematician (though he became the Dauphin's tutor), but he called in some of the most

[1] Turnbull *op. cit.* pp. 286, 288–9, etc.; Hooke, *Diary, 1672–80* (ed. H. W. Robinson and W. Adams, London, 1935), 11, 17, 23 September 1674.

able of the Académie to assist him.[1] In the first part of the *Art de Jetter les Bombes* he discusses the ballistical ideas which had prevailed before Galileo proved that the trajectory of a projectile in a vacuum is a parabola. He then goes on to discuss the calculation of the horizontal ranges of cannon in accordance with the parabolic hypothesis, either by means of the table of sines, or by means of the improved gunner's quadrant invented by Torricelli. Next he approaches the calculation of ranges when the mouth of the gun and the target are not in the same horizontal plane, giving first Torricelli's method of calculating ranges over sloping ground when the angle of elevation of the piece is known, and then the converse of Torricelli's problem, that is, the way of calculating the proper angle of elevation in order to hit a target at a known distance upon a sloping plane.[2] Four different geometrical methods are proposed and the use of various instruments for the same purpose described.

The third part of the book, amounting to a full exposition of the writings of Galileo and Torricelli on the theory of projectiles, elaborates the mathematical theory of the methods and instruments suggested for handling the problems encountered in gunnery. The geometrical calculations for the angle of elevation required to shoot over a given plane provided by Milliet de Challes (1621–78), an encyclopaedic writer on mathematics, Jacques Buot, mechanician, Ole Roemer (1644–1710), the Danish astronomer who was then in Paris, and Philippe de la Hire (1640–1718), a notable mathematician and like the others a member of the French academy, are examined in detail. There is also a description of the method developed by Jean Cassini (1625–1712), the famous astronomer who discovered the division in the ring of Saturn, for solving the same problem, and of the universal instrument for gunnery constructed in accordance with it.

The fourth and final part of the book treats of the difficulties to be overcome by ballisticians, placing among these the resistance

[1] The book was written in 1675, the dedication being dated Jan. 1676, but not published until 1683 because it was feared that Blondel's doctrines might assist the enemies of France. Blondel remarks, 'Cest donc, Sire, pour seconder, en ce que je puis, de si glorieux desseins, que je mets ce Traité au jour'.

[2] I.e. $\frac{\sin (2a-b)}{\sin b} = \frac{1 + gR \cos b}{v^2 \tan b}$, where a is the angle of elevation, b the angle of the slope R, the range required, v the initial velocity.

of the air. Blondel concludes that it is impossible to calculate exactly the effect of this resistance upon the trajectory of the projectile, but that for round and heavy bodies such as cannon-balls and bombs the alteration is not great: 'l'on ne doit pas presumer que la résistance de l' air apporte de grands changemens dans les mouvemens de nos projections ordinaires.'[1] Moreover, the proportions of various ranges of the same gun may be correct within themselves, though each one differs slightly from that expected according to the parabolic theory, and the effect of resistance in shortening the trajectory by reducing the horizontal component of velocity is countered by its effect in prolonging the trajectory through equally resisting the vertical gravitational component.[2] Having answered all the objections which may be opposed against the parabolic theory, Blondel proceeds to quote many experiments made by the Academicians with fountains of water and quicksilver, and ballistic machines which confirm it.[3] He insists on the real usefulness of ballistics in war, adding that his opinion was fully shared by Louis XIV, who had had the Dauphin instructed in the mathematical parts of warfare.

Although Blondel's book was a most copious treatise on the theory of gunnery, there was nothing novel in it; containing some good work on the application of mathematics, as an essay in scientific ballistics it was already out of date, as the number of references to Mersenne (whose experiments Blondel was unable to appreciate properly) prove. The same may be said of Halley's essays in the *Philosophical Transactions* for 1686-7 and 1695-7, in which he offered an algebraic solution to the problem of finding the angle of elevation when the ground is not horizontal. The most acute mathematicians were no longer interested in the parabolic theory, which had been fully worked out, and were turning to the analysis of air resistance. Practical gunners, reluctant to renounce the simple system for their art proposed by Galileo, and incapable of understanding the subleties of such men as Huygens, Gregory, Wallis and Newton, valued the books of Blondel and Anderson.[4] The writers who have rightly dwelt on Galileo's

[1] *Op. cit.* p. 345 (Paris, 1683). [2] *Ibid.* p. 358.

[3] These were principally made by Perrault and Roemer.

[4] E.g. John Harris, in his *Lexicon Technicum* (1704) article on 'Ordnance', follows Anderson and Halley, with no mention of the resistance of the air.

achievements in theoretical mechanics, ignoring the practical mind of the seventeenth century and its tendency to look upon the problems of science as the problems of real nature, have sometimes misunderstood the criticisms to which Galileo's theories were exposed. The physicist, with the lingering tradition of the medieval philosopher in him, sought for something beyond a hypothesis involving abstractions and units which he could neither see nor measure. He preferred certain, mechanical or material explanations. He chose rather to think in terms of material corpuscles than of atoms and waves, of vortices than of attractions.

The ballistical problem was more than a question of the application of science to technology. It was a part of the philosophical crisis which had been apparent since Galileo had rejected Aristotle's explanation of all natural phenomena by a combination of common-sense observation: the difficulty of reconciling theory and experiment. Scientific theories and scientific laws would remain unconvincing unless they could be made to give accurate results in matters of detail. By the third quarter of the century everyone in the van of the scientific movement admitted that the primary principles of dynamics laid down by Galileo were fundamental to all future work, but it was also apparent that in their simple form they were not true for the world of experience. Bodies do not move eternally along straight lines, nor do falling bodies accelerate exactly according to Galileo's law—even if the rotational deviation discussed by Hooke and Newton is left out of account. The same difficulty existed in astronomy, where the detailed work of applying the new cosmological theories in detail —for example to the period of the moon—had hardly been begun. It was necessary to discover the complex mathematical rules which link the world of scientific abstraction to the world of nature, if it was to be proved that the one was indeed appropriate to the other—in other words to begin the work which Newton attempted in the *Principia*, for, as he says in his Preface, 'the whole burden of philosophy seems to consist in this—from the phenomena of motions, to investigate the forces of nature, and then from these forces to demonstrate the other phenomena', down to the least detail.

The confusion of the two fields of pure science and engineering, with all their different needs and problems, almost complete in Leonardo da Vinci, for example, was lessening but had not yet altogether disappeared. There was still a strong strain of practical mechanics in Huygens, Hooke, Newton and many more. They were keenly interested in refining science until it was capable of dealing with mundane as well as sublime problems. When Newton undertook to clear up the whole problem of resistance in the second book of his *Principia* he was guilty neither of an aberration in scientific work nor of allowing himself to be dominated by the technological problems of his own day. He was using modern mathematics in order to write what was, in a sense, the last word on a philosophical question over which philosophers had puzzled for centuries.

CHAPTER VI

MATHEMATICAL BALLISTICS III

Fluxions and Calculus

The publication of James Gregory's mathematical papers has shown that he also was rapidly extending the infinitesimal calculus which in the hands of Newton developed into the method of fluxions; Gregory suggested the reflecting telescope which Newton perfected into a practicable instrument; and in the science of ballistics it was Newton who substituted a sound and splendid analysis for the few propositions that Gregory had sketched. This parallelism of their works points to a common interest in some of the dominant problems of physico-mathematical science. About fourteen years elapsed before John Wallis published a paper containing the first reinvestigation of the question raised by the Scottish professor, though it was already known (by December 1684, when Halley brought the news to the Royal Society) that Newton, who had been appointed Lucasian Professor of Mathematics in 1669, had recently given a series of important and original lectures on motion which included a discussion of the effect of a resisting medium upon the motion of a particle.[1] The announcement apparently stimulated Wallis to activity, since two years later Halley, who was responsible for the business of the Royal Society, as curator of experiments and publisher of the *Philosophical Transactions* wrote to him:

You were pleased to mention some thoughts you had of communicating your conclusions concerning the opposition of the Medium to projects moving through it; the Society hopes you continue still inclined to do so, not doubting but that your extraordinary talent in matters of this nature, will be able to clear up this subject which

[1] More, *op. cit.* pp. 299 *et seq.*; MacPike, *op. cit.* pp. 6-7. Newton's lectures for 1684-5 (representing part of bk. I of the *Principia*) are in the Cambridge University Library, with drafts of ballistical propositions. His lectures for 1686, in which he may have continued with bk. II, are lost.

hitherto seems to have been only mentioned among Mathematicians, never yet fully discussed.[1]

With this flattering invitation to instruct the premier scientific assembly of Europe he enclosed a copy of Newton's propositions on the same subject, which he was well able to do as he had undertaken to see the *Philosophiae Naturalis Principia Mathematica* (as the book which grew out of the lectures on motion was to be called) through the press. Wallis was at first disinclined to anticipate the publication of Newton's book, since he found that Newton had saved him the

> labour of doing the same thing over again. For I should have proceeded upon the same principle that the resistance (caeteris paribus) is proportional to the celerity (because in such proportion is the quantity of air to be removed in equal times) nor do I know from what more likely principle to take my measures therein.[2]

In spite of this diffidence Halley continued to urge Wallis to pursue his inquiries and particularly to examine whether

> the opposition (of the medium) may not take off, constantly such a part of the velocity in each part of the motion, as the intrinsick or specifick gravity of the medium is of the gravity of the body projected; which if true will give us a better account than can be obtained from experiments,[3]

a suggestion which clearly harks back to Galileo's ideas. Finally Wallis was persuaded to publish a paper to 'adorn or recommend' the *Transactions* in the spring of 1687, an argument, Halley wrote, 'that will most recommend my collection to the curious, since so difficult and nice a subject is so clearly and throughly handled therein'.[4]

From Wallis' own admission, it will appear that his paper was considerably inferior to Newton's investigation, published a few months later, and to those of Huygens, still unknown to anyone; and indeed it seems that it was only when Wallis had agreed to publish that Halley informed him that Newton had gone on to

[1] 11 Dec. 1686 (MacPike, *op. cit.* p. 74).

[2] Wallis to Halley, 14 Dec. 1686 (Birch, *op. cit.* vol. IV, p. 514).

[3] Halley to Wallis, 15 Feb. 1686/7 (MacPike, *op. cit.* p. 80).

[4] 9 April 1687; 25 June 1687 (MacPike, *op. cit.* pp. 81, 85). Obviously Halley, who was not noted for tact, acted with very little discretion in pressing another mathematician to write upon a problem taken from a manuscript then in his own hands, and actually communicating to Wallis Newton's own reasoning before it was published.

consider the hypothesis of the resistance being proportional to the square of the velocity. Wallis simply treated it as increasing directly with the velocity of the moving body, finding an experimental demonstration of resistance in the fact that a cannon-ball projected horizontally does not strike so hard against a wall at a great distance as against one close at hand. Suppose, then, the 'Force impressed (and consequently the Celerity if there were no resistance)' as 1, the resistance as r (which must be less than one or no motion could take place), the effective force in the first moment will be $1-r$. Let $m=\frac{1}{1-r}$; in the first instant 'the effective force (and consequently the Celerity)' is $\frac{1}{m}$ of what it would have been without resistance, in the second $\frac{1}{m^2}$, in the third $\frac{1}{m^3}$, etc. Now since the distances passed over are proportional to these velocities and forces, they are decreased in the same way, and since these fractions form a decreasing geometrical progression, if it is infinitely continued 'determining in the same point of Rest where the motion is supposed to expire' its sum is of finite magnitude, and the distance passed over is $\frac{1}{m-1}$ 'of what it would have been in so much Time, if there had been no resistance'.

Wallis goes on to show how the progression can be applied geometrically by means of the hyperbola, but adds that the 'proportion of r to 1 or (which depends on it) of $1-r$ to 1, or 1 to m, remains to be inquired by experiment'. Having discovered in this way how far a body will penetrate into a resisting medium when projected horizontally, he proves by similar reasoning that if it is projected vertically downwards it descends 'at the rate $\frac{1}{m}$, $\frac{1}{m^2}$, $\frac{1}{m^3}$, etc. of f (the impressed force) increased by $\frac{1}{m}$, $\frac{1}{m}+\frac{1}{m^2}$, $\frac{1}{m}$ $\frac{1}{m^2}+\frac{1}{m^3}$, etc. of g the impulse of gravity', while if projected vertically upwards it ascends at the former rate diminished by the latter since now the force of gravity is opposed to that of projection. As the motion of a projectile may be resolved into horizontal and vertical components, its trajectory may be determined from these relations, and Wallis accordingly concludes that 'In a Horizontal or Oblique projection; if to a Tangent who's increments are . . . as $\frac{1}{m}f$ &c, be fitted Ordinates (at a given angle) who's increments are as . . . $\frac{1}{m}g$ &c, the Curve answering to the compound of these Motions is that wherein the Project is to move', and suggests that this curve, resembling a parabola deformed, should be known as the *linea projectorum*.[1]

Wallis's discussion of resisted motion was marred not only by his confusion of velocity and force, but by his very clumsy mathematical method which denied him any precise conception of the curve he was considering or of its derivation from other known curves, and when the *Principia* appeared the sections devoted to the same subject were incomparably finer than anything that had been done before. Even Huygens, who alone could have prided himself on a priority over part of Newton's work, freely admitted that Newton had outstripped all other philosophers in his study of the resisting medium. It was most striking in its masterly completeness. Ballistics was only an incidental in a wide survey that included an investigation of the propagation of light and sound, as well as of the circular motions of bodies. Whereas previous writers had at best discussed resistance in a few paragraphs as a source of inconvenient modifications to natural laws, Newton's study was divided into forty propositions occupying 126 pages in the third edition of the *Principia*, in which he examined the consequences which ensued according to mathematical reasoning from different types of motion in mediums presumed to resist in proportion to the velocity, the square of the velocity, or these two conjointly.

This was only made possible by the use of the method of fluxions briefly described in Lemma II, which was the immediate instrument of his interest in working out these theorems as well as of his success in resolving them; for while the lack of an appropriate mathematical technique had stifled the curiosity of Huygens beneath a burden of computation, to Newton each successive

[1] 'A Discourse concerning the Measure of the Air's resistance to bodies moved in it', *Phil. Trans.* no. 186, Jan.-March 1686/7, pp. 269 *et seq.*

proposition was a revelation of the powers of his new analysis.[1] As he suggested in the scholium to Prop. XIV, there were many other ways in which the effects of resistance upon moving bodies might be analysed, in addition to those he had touched upon, but he 'hastened to other things' after dealing with those most suitable for his purpose.[2]

No one before Newton had attempted to prove, by detailed consideration of circular motion in a resisting medium, that it was essential to the stability of the planets (in the idea of the universe then admitted by science) that in their rotation about the sun they must encounter infinitesimal resistance or none, a fact which he took to be a cogent argument against the Cartesian system. No one, though many vague suggestions had been put forward, had calculated the influence of air resistance upon the oscillations of pendulums, or ventured to predict the shape which would be most favourable to motion in a resisting medium. It is not often remembered that the only experimental researches of any note reported in the *Principia* are in connection with the discussion of the resistance of air and water, for Newton was the first to make experiments on the times of fall of bodies of different densities over a considerable distance, measured with the greatest possible accuracy, and to demonstrate that his theory was in very close agreement with the results of observation, not in one instance but in many. In fact he added fresh experimental proof of the justice of his method to each revision of the *Principia*, using in the second edition the experiments of Hawksbee, in the third those of Desaguliers.

Why did Newton, unlike other philosophers of the time, expend so much effort upon the theory of resistance, both in

[1] *Principia*, p. 243. The references are to the third edition, London 1726, and English translations are taken from the version of Florian Cajori, California, 1934. Cf. *Phil. Trans.* (1715): 'By the help of the new Analysis Mr Newton found out most of the Propositions in his *Principia Philosophiae* but because the ancients for making certain things admitted nothing into geometry before it was demonstrated synthetically, he demonstrated the propositions synthetically that the system of the heavens might be founded on good geometry. And this makes it now difficult for unskilful men to see the Analysis by which these propositions were found out' ('An Account of the Book entitled Commercium Epistolicum', p. 206). This long account of the history of fluxions was written (or at least inspired) by Newton himself. A longer version of the passage quoted, in Newton's hand, is in Add. 3968. Bundle 13, Cambridge University Library.

[2] *Principia*, p. 274.

vi secundi horis quarti &c. Proinde cum areae aequales BKx, KLλ, λLMμ, μMNν &c sint viribus centripetis analogae, erunt rectangula Bkx, kLλ, λMμ, μNν &c resistentiis in medio singulorum temporum, hoc est celeritatibus atque adeo descriptis spatiis analogae. Sumantur analogarum summae et erunt areae Bkx, Blλ, Bmμ, Bnν &c spatiis totis descriptis analogae, nec non areae ABxK, ABλL, ABμM, ABνN &c temporibus. Corpus igitur inter descendendum, tempore quovis ABλL describit spatium Blλ, et tempore Lλμν spatium λlnν. Q.E.D. Et similis est demonstratio motus expositi in ascensu. Q.E.D.

Schol. Beneficio duorum novissimorum problematum innotescunt motus projectilium in aere nostro, ex hypothesi quod aer iste similaris sit quodque gravitas uniformiter & secundum lineas parallelas agat. Nam si motus omnis obliquus corporis projecti distinguatur in duos, unum ascensus vel descensus alterum progressus horizontalis: motus posterior determinabitur per problema sextum, prior per septimum ut fit in hoc diagrammate.

Ex loco quovis D ejaculetur corpus secundum lineam quamvis rectam DP, & per longitudinem DP exponatur ejusdem celeritas sub initio motus. A puncto P ad lineam horizontalem DC demittatur perpendiculum PC, ut et ad DP perpendiculum CI, ad quod sit DA ut est resistentia medii ipso motus initio ad vim gravitatis. Erigatur perpendiculum AB cujusvis longitudinis et completis parallelogrammis DABE, CABH, per punctum B asymptotis DC, CP describatur Hyperbola secans DE in G. Capiatur linea N ad EG ut est DC ad CP et ad rectae DC punctum quodvis R erecto perpendiculo RT quod occurrat Hyperbolae in T et rectae EH in t, inde cape $Rr = \frac{DRtE - DRTBG}{N}$ et projectile tempore DRTBG perveniet ad punctum r, describens curvam lineam DarFK quam punctum r semper tangit, perveniens autem ad maximam altitudinem a in perpendiculo AB, deinde incidens in lineam horizontalem DC ad F ubi areae DFsE, DFSBG aequantur et postea semper appropinquans Asymptoton PCL. Estque celeritas ejus in puncto quovis r ut Curvae Tangens rL.

Si proportio resistentiae aeris ad vim gravitatis nondum innotescit: cognoscantur (ex observatione aliqua) anguli ADP, AFr in quibus curva DarFK secat lineam horizontalem DC. Super

NEWTON'S FIRST SOLUTION OF THE BALLISTICAL PROBLEM, 1684

calculation and in experiment? To this question there is no single convincing answer, but he threw some light upon it in his own account of his aims and methods. He wrote, at the opening of the Third Book:

In the preceding books I have laid down the principles of philosophy; principles not philosophical but mathematical; such, namely, as we may build our reasonings upon in philosophical inquiries. These principles are the laws and conditions of certain motions, and powers and forces, which chiefly have respect to philosophy; but lest they should have appeared of themselves dry and barren, I have illustrated them here and there with some philosophical scholiums, giving an account of such things as are of more general nature, and which philosophy seems chiefly to be founded on; such as the density and the resistance of bodies, spaces void of all bodies, and the motion of light and sounds.

He went on to explain that his first idea for his Third Book on the system of the world was to write it in popular language, *ut a pluribus legeretur*, but on maturer consideration he resolved to reduce it to mathematical propositions which could only be understood by those who had first mastered the earlier books, and were thus prepared to appreciate the climax of his work, in order to avoid the popular prejudices which might arise in favour of older notions. And he advised the reader of the third book to peruse carefully 'the Definitions, the Law of Motion, and the first three sections of the first book. He may then pass on to this Book, and consult such of the remaining Propositions of the first two Books, as the references in this, and his occasions, shall require.'[1]

Newton knew what all later commentators have admitted, that what he had to say on the structure of the universe was vastly more important in the history of thought than his contributions, however interesting and novel, to physical and mechanical science. Most of his study of physics was only by way of illustration of the mathematical principles which he believed to operate throughout the universe, and this he advised the student who was looking for his fundamental ideas to omit. His work on ballistics, on the flowing of fluids, on pendulums, on light and sound, was, like everything Newton thought worth printing, of supreme value,

[1] *Principia*, p. 386. This in fact was how the *Principia* was studied in Cambridge in the eighteenth and nineteenth centuries.

but it was incidental to the main object of the *Principia*, the interpretation of the universe as a whole.

The question must next be asked, what was Newton's purpose in writing the two first books at such length, when, as he remarked, 'they abound with such as might cost too much time, even to readers of good mathematical learning'? The difficulty here is that although we have the internal evidence of the *Principia* before us, Newton left very little external evidence of the course of his thought. Almost complete obscurity hides the progress of his early studies and the influence of his teachers. We know that he had carefully examined Descartes' *Principia Philosophiae*, but it does not appear from surviving lists of his books that he possessed, for instance, a copy of Galileo's *Discourses*, or anything of Huygens except *Horologium Oscillatorium*; Mersenne's name is absent from them, so too is that of Torricelli.[1]

His extant correspondence is of little more use for this purpose; he wrote eight letters to Halley in the years 1686–87, and in only one of them refers to his theory of projectiles.[2] It can be concluded tentatively that Newton cast his first two books in this form partly because he wished to satisfy the reader in problems of detailed physics as well as in the grand structure of the universe, partly because his principles deserved some preliminary display before they were imposed on the reader in a way which might well challenge his credulity, while in them undoubtedly survived vestiges of a treatment more appropriate to the academic lecture than to the philosophy Newton constructed from the unfolding of his ideas. He knew that he could add much to contemporary knowledge in physics and mechanics, without composing a physical system like that of Rohault and the other followers of Descartes, by including some of his meditations on the disputed points of seventeenth-century physics as they were relevant to his purpose. Occasionally, as in the thirty-seventh proposition of

[1] R. de Villamil, *Newton: the Man* (London, n.d.), appendix. See also my paper published in the *Cambridge Historical Journal*, vol. IX (1948).

[2] 18 Feb. 1686/7. '[Mr Montague] writes that Dr Wallis has sent up some things about projectiles pretty like those of mine in the papers Mr Paget first showed you, and that 'twas ordered I should be consulted whether I intend to print mine. I have inserted them into the beginning of the second book with divers others of that kind, which therefore if you desire to see, you may command the book when you please' (W. W. Rouse Ball, *An Essay on Newton's 'Principia'* [London, 1893], p. 169).

Book II in the first edition, reliance on his principles caused him to fall into an error which a simple experiment would have exposed.[1] Newton's interest in the technological application of his mathematical principles was slight, though not altogether absent. When he remarked in the scholium to Prop. 34 of Book II, 'This proposition I conceive may be of use in the building of ships', it was only to throw out a suggestion that he did not trouble to pursue, and such an aside should not be construed as though it were typical of the mood in which the *Principia* was written. Similarly the propositions on ballistics were almost entirely abstract as a consequence of Newton's approach to them. He sought for the solution of a mathematical problem; whether or not that solution was useful was of little more concern to him than whether his experiments on light fitted in with prevailing theory.[2]

Ballistics only appears in the *Principia* in the solution of general equations for the movement of a mass-point resisted in various hypothetical ways. Newton knew nothing of guns, he carried out no experiments in gunnery. Only the 'laboratory' work, the experiments on the acceleration of falling bodies in air, attracted him.[3] For this he was able to offer a precise formula.[4] Nor did

[1] For further light on this and on seventeenth-century scientific method—which does not at all comply with traditional views—see Halley's remarks to the Royal Society on this point (MacPike, *op. cit.* pp. 147–9, 222).

[2] Rouse Ball prints the following passage, which I translate, from the 'De Motu sphaericorum Corporum in fluidis', one version of which was referred to by Newton (p. 136, n. 2 above). This MS. was probably written in 1685; in the scholium after the Vth Problem Newton writes: 'moreover the motions of projectiles in our air are to be referred to the immense and indeed motionless sphere of the heavens, not to the moving space which revolves as one with the earth and our air, and is by the untaught looked upon as motionless. The Ellipse is to be found which the projectile describes in this really motionless space, and then its motion in the moving space determined. When this is settled it may be deduced why the heavy body, falling from the top of a lofty building, is deflected a little in falling from the perpendicular, how great this deflection is and in what direction it occurs. And conversely from the deflection the proof of the rotation of the earth may be deduced by experiment. When I myself formerly signified this deflection to the celebrated Hooke he confirmed by an experiment three times repeated that it was so, the body being always deflected from the perpendicular towards the east and south, as is correct in our northern latitudes' (Rouse Ball, *op. cit.* p. 56). Thus in Newton's mind the ballistical propositions of Book II of the *Principia*—expanded from this MS.—were intimately linked with the problems of astronomy and gravitation.

[3] *Principia*, pp. 344 *et seq.* The only instance I recollect of Newton taking any cognisance of a practical question of artillery is a note by David Gregory that 'Sir Isaac Newton believes that a mixture of Copper, Tin, Constantinople Spelter, and a little silver will doe [for the rebushing of touch-holes], if while this is a running into the breach or wide hole, the gun be very hot'. (1705? W. G. Hiscock, *David Gregory, Isaac Newton and their Circle*, Oxford 1937, p. 25.)

[4] *Principia*, pp. 344, 355.

he maintain that any of the hypotheses of resistance he used was physically true, or that the mathematical demonstrations based upon them corresponded exactly with the path of a real projectile, thus avoiding the mistake of Galileo and Torricelli of confounding geometrical and empirical truth. He did not go beyond the guarded statement that the

> resistance of bodies in the ratio of the velocity is more a mathematical hypothesis than a physical one. In mediums void of all tenacity, the resistances made to bodies are as the square of the velocities.[1] For by the action of a swifter body, a greater motion in proportion to a greater velocity is communicated to the same quantity of the medium in a less time. . . .

After discussing the hypothesis that the resistance of the medium is partly directly proportional to the velocity of the moving body, partly proportional to the square of the velocity he continues:

> The resistance of spherical bodies in fluids arises partly from the tenacity, partly from the attrition, and partly from the density of the medium. And that part of the resistance which arises from the density of the fluid is, as I said, as the square of the velocity; the other part, which arises from the tenacity of the fluid is uniform, or as the moment of time.

Other mathematical hypotheses different from these might be framed; none of them, especially the first two, were actually true of a physical medium. To refer to the hypothesis $R \propto V^2$, as 'Newton's law of resistance' is therefore an unjust simplification of his words.[2]

The only ballistic curve which he was able to solve completely was the simplest and least probable, on the hypothesis of the resistance being directly proportional to the velocity (Book II, Prop. IV). To avoid excursions into tedious geometrical detail (which is of course readily available in Florian Cajori's recent edition of the *Principia*, and in the notation of the calculus in

[1] 'Caeterum, resistentiam corporum, esse in ratione velocitatis, hypothesis est magis mathematica quam naturalis. In mediis, quae rigore omni vacant, resistentiae corporum sunt in duplicata ratione velocitatum' (*Principia*, p. 239). In the first edition Newton wrote: 'Obtinet haec ratio quam proxime ubi corpore in mediis rigore aliquo praeditis tardissime moventur. In mediis autem quae rigore omni vacant (uti posthoc demonstrabitur) corpora resistantur in duplicata ratione velocitatum' (1687, p. 245).

[2] *Principia*, p. 274.

various publications), I shall discuss Newton's methods and results very briefly. He introduces resistance in a ratio to the acceleration of gravity, $\frac{x}{g}$, and gives a method of finding both this ratio and the initial velocity of projection by interpolation from known ranges. The curve which is the path of the projectile through the air he constructs by drawing ordinates to a rectangular hyperbola which is asymptotic to the horizontal plane and to a perpendicular, which is also the vertical asymptote of the trajectory. This is in fact the curve obtained by taking abscissae from the point of origin in decreasing geometrical progression and corresponding vertical ordinates from an oblique axis (which is the tangent to the trajectory at the point of origin) in increasing arithmetical progression. Thus the general characteristic of the trajectory is greater steepness in the descending than in the ascending branch, and as Newton says, it may be 'easily delineated by the table of logarithms'.[1]

To find an equation for the trajectory in a medium resisting as the velocity was a relatively simple task, but Newton found that if the resistance was supposed to increase as the square of the velocity he was not able to arrive at the same happy issue, though like Huygens he was able to solve for the times and distances of vertical ascent and descent (Props. VIII and IX). After some necessary preliminaries, including a short notice of his method of fluxions, he shows that, following this hypothesis, the trajectory can be neither a parabola, a circle, nor a hyperbola, using these curves of course solely in order to demonstrate his method, not because one of them might be the one required.

It is obvious that if the resistance of the medium is proportional to the square of the velocity of the projectile and the density of the medium, and the density is supposed to vary along the whole path of the projectile in accordance with some law, it is possible to find the law of the variation of density necessary to produce any given curved trajectory. If a number of different curves are tested in this way and the appropriate laws of variation ascertained, any curve which gives a *uniform* density along the whole trajectory may be that which a projectile really follows—in accordance

[1] *Principia*, pp. 234 *et seq.*, Corollary II (added in the second edition, 1713), Corollary VII. Cf. Newton's letter to Collins, June 1674, quoted above, p. 123.

with the selected hypothesis of resistance. This reversal of the normal procedure which Newton employed for the first time has been called the inverse problem of ballistics, and by this means he proved that the density in the case of a parabolic trajectory is zero (Galileo's well known hypothesis), that in the case of a circular trajectory the variation was from positive to negative, and for the hyperbola the density must have a continuous positive variation along the path of the projectile.[1]

Although some continental mathematicians at a later time were greatly elated at the discovery of some oversights of which the English philosopher was guilty in the calculations of this proposition, and tried to use them to prove that Newton did not even understand his own fluxions, these mistakes in no way affected the result. Of the curves he considered the hyperbola appeared to offer the closest approximation to the curve he sought. Abandoning the consideration of this highly complex trajectory, therefore, he sums up his investigations in a renunciation of conventional ballistics and advises the use of empirical methods:

> Since there can be no motion in a parabola except in a non-resisting medium, but in the hyperbolas here described it is produced by a continual resistance; it is evident that the line which a projectile describes in a uniformly resisting medium approaches nearer to these hyperbolas than to a parabola. That line is certainly of the hyperbolic kind but about the vertex it is more distant from the asymptotes, and in the parts remote from the vertex draws nearer to them than these hyperbolas here described. The difference, however, is not so great between the one and the other but that these latter may be commodiously enough used in practice instead of the former. And perhaps these may prove more useful than a hyperbola that is more accurate, and at the same time more complex.[2]

[1] Bk. II, Prop. X (*Principia*, pp. 252 *et seq.*). An error in the lengthy and difficult mathematics of this proposition extending over twelve pages, first pointed out by Nicholas Bernoulli who visited England in the autumn of 1712, and accordingly signified to Roger Cotes, at that time Plumian Professor and engaged in the preparation of the second edition of the *Principia* for the press, made it necessary for the demonstration of Prop. X to be completely revised. Newton had made the necessary corrections by Jan. 1712/3. As Cotes then remarked, the only effect of the change was to increase the resistance in the cases of the circle and the hyperbola in the ratio of three to two (Edleston, *op. cit.* pp. 142-6). The existence of an error undetected for a quarter-century is sufficient evidence both of Newton's lead over other mathematicians and of the obscurity of his methods.

[2] *Principia*, p. 260.

Newton then goes on to give eight rules which determine the properties of a hyperbola representing accurately enough, as he thought, the trajectory in air, and from these describes a construction by which the hyperbola can be traced from two observations.

Newton's predecessors had always considered the projectile as a small solid sphere of weighty material which could be treated mathematically as a mass-point, and he created the 'problem of the solid of least resistance' by discussing for the first time the effect of the shape of a moving body upon the resistance experienced. The problem amounts to finding the curve which shall produce a solid of revolution of given diameter having less air-resistance than any other solid of the same diameter. Shipbuilders had of course tried very crudely to design their craft to slide smoothly through the water and to obtain the highest speed with the least force, but the use of any other than round-shot was not feasible in the artillery of Newton's time, so that his problem was not a practical one. It was, however, one that might easily arouse the interest of a consummate mathematician since it demanded the greatest proficiency in the calculus—or fluxions—and in fact Newton is considered to have inaugurated the calculus of variations, in this proposition. Professor Turnbull has stated that Newton's solutions place him 'in relation to this new subject in the same pioneering rank as Fermat in relation to the differential calculus'.[1] Because he was only able to prove his results by the method of fluxions and not by the synthetic method adopted throughout the *Principia*, he omitted the demonstration from his proposition,[2] in which he reaches the apparently anomalous result that the resistance is least, not when the solid has a smooth ogival shape presenting a point in the direction of motion, but when it ends in a small flat face, i.e. the curve of the solid of least resistance does not meet the axis of the solid.[3] Later mathematical researches have revealed that the requirements for a true minimum are more complicated than Newton knew, and that in fact the problem as stated by Newton, on the hypothesis of the resistance being

[1] H. W. Turnbull, *Mathematical Discoveries of Newton* (London, 1945), p. 39.

[2] The proof is available in a letter in the Portsmouth Collection, probably written to David Gregory of Oxford, printed in the *Catalogue* to the Collection and in Cajori's edition of the *Principia*, pp. 657–9.

[3] *Principia*, pp. 323, *et seq.*

proportional to the square of the velocity, is incapable of a perfect mathematical solution.[1]

As this last result suggests, the effect of Newton's studies in the *Principia* was twofold; to show that the methods of simple mathematics were no longer adequate for the theoretical treatment of the problems of ballistics, and that these fell into two clear divisions. The experimenter had to discover how to measure the velocities of projectiles and the resistance they experience in the air; the mathematician had to equip himself to calculate trajectories under given physical conditions, in which he would find himself deprived of any easy working hypothesis. Whatever the physical properties of the resisting medium, Newton had shown that the mathematician's task was extremely difficult.

Yet it was the purely mathematical work that was more successfully carried on in the generation after the publication of the *Principia*. Proper experimental foundations for the study of ballistics were long lacking; Robins' observations on the velocity of projectiles and air resistance with the ballistic pendulum were not made known until 1742, and for obvious reasons in his experiments he was limited to the use of small projectiles and charges. Late in the century his methods were extended by Hutton, with official facilities, to cannon, and in the next progress in electricity made the invention of the Bashforth chronograph possible. But in mathematics, where Newton cleared a broader path, others of sufficient competence to push forward were more ready to follow his lead. This of course was only possible with the aid of a calculus at least equal to his own in power and superior in convenience. Consequently the importance of the Second Book of the *Principia* is greater in the history of mathematics than in the history of technology, and the further essays on ballistics which it provoked belong not so much to the seventeenth as the eighteenth century, to the period of the development of the Leibnizian calculus. The mathematicians of a younger generation, such as Varignon and the Bernoullis, had no more practical interest in ballistics and gunnery than the seventeenth-century scientist. Something much more fundamental to science was at stake, the

[1] (Cajori) *op. cit.* p. 661; A. R. Forsyth, 'Newton's Problem of least Resistance' (in W. J. Greenstreet, ed., *Isaac Newton: a memorial volume*, London, 1927).

translation of the existing body of dynamical knowledge into the notation of the calculus, bringing about not only the elaboration of that great instrument, but the deepening of knowledge in the mechanical sciences themselves. The calculus, or method of fluxions, was the mathematics of changing quantities, lending itself readily to the interpretation of problems of motion, and the older methods, including those of the *Principia*, were soon swept away, on the Continent at least.

The papers on ballistics which appeared in the scientific journals between the first and third editions of the *Principia*, belonging to the eighteenth century and a new era of mathematics, will only be briefly mentioned here for their interest as reflections of Newton's work. It must be recalled that at the end of the century English science was at a low ebb. The Royal Society was in debt, its meetings thinly attended, for court and country lost interest in it as politics at home and abroad occupied every thought. Men were proud of the *Principia* without understanding it, and Newton had few pupils. The great figures of the renaissance of science in England were dead or had passed on to other occupations, like Wren (and Newton himself after 1696); John Wallis, the finest English mathematician after Newton, was too old to learn new ways.

As a result the extension, soon degenerating into a blind defence, of Newton's great advances in mathematics rested with mediocrities like Keill and Raphson, who confused science and national egotism. The best mathematical writing came from the pens of continental mathematicians, of whom it must be said that they were no better mannered than their English adversaries.[1] In the course of the bitter disputes over the invention of the differential calculus, which divided England from the main stream of European mathematics for over a century, valuable work was done less in the interest of the advancement of mathematics than of exposing the ignorance and ethical weakness of Newton's supporters, while these in turn suspected and belittled everything that came from abroad. Equations became the instruments of passion

[1] David Gregory notes that Johann Bernoulli 'is a very rough, rude man, speaking well of no mortall, and giving names to everybody' (W. G. Hiscock, *David Gregory, Isaac Newton and their circle*, Oxford, 1937).

and prejudice. In so far as the mathematics of gunnery became involved in the controversy, the continent had the advantage through the use of Leibniz's calculus, almost unknown in 1687 but familiar by 1700.

First remarks on the theory of resistance came from Leibniz himself in an article entitled 'Schediasma de Resistentia Medii, & Motu Projectorum gravium in medio resistente' which appeared in the periodical founded by him, the *Acta Eruditorum*.[1] From the neglect of the resistance of the air by Galileo, Torricelli and Blondel, he wrote, arose the discrepancies between theory and observation in gunnery. Yet the true laws of projectiles, and a calculus agreeing with experiment, things which would be of the greatest use in ballistics and pyrobolics, depended directly upon these matters. He claimed to have already communicated some thoughts on this subject to the Académie des Sciences when in Paris.[2] Resistance he divided into two species: absolute resistance is a uniform reduction of the force of the moving body irrespective of its velocity, caused by the viscosity of the medium, comparable to friction, and is proportional to the area or contacting surface of the body; respective resistance is caused by the density of the medium, and is proportional to the velocity and volume of the moving body, because in this ratio are the particles of the medium more violently agitated. Therefore if a body accelerates through a medium having respective resistance, its velocity cannot increase beyond a certain limiting value.[3] Leibniz next considered the shape of the trajectory of a projectile in a medium resisting in either of these ways as a combination of a uniform retarded motion with an accelerated retarded motion. In each case he indicated, though he did not explore, methods of calculating the coordinates of any point P which is the locus of the projectile from the known relations of uniform and accelerated

[1] January 1689, pp. 38–49, printed in Gerhardt, *Leibnizens mathematische Schriften*, vol. VI, p. 135.

[2] 'Ego jamdudum Inclytae Academiae Scientiarum Regiae Parisinae, cum apud illos agerem de hoc argumento ratiocinationes communicavi, & modum aestimandi ex parte tradidi, speciesque distinxi.'

[3] '. . . a resistentia vero respective corpus uniformiter acceleratum (ut grave descendens) habet certum limitem velocitatis, seu maximam velocitatem exclusivam, ad quam semper accedit, (ut postremo differentia sit insensibilis) it tamen ut eam nunquam perfecte attingit.'

resisted motion. To combine the two results of a medium having both absolute and respective resistance, such as occurs in nature, Leibniz thought to be by no means impossible, and that much of use might follow from such a calculation, but stated that he himself was only concerned to give the geometrical elements of the method.[1]

A little later Huygens entered into the discussion. His first news of an important book on philosophy by Newton, whose optical experiments he had criticised in 1672, came from Fatio de Duillier, a young acquaintance of his who had taken up residence in England and become a keen partisan of Newton, a month before the *Principia* was published.[2] Probably Huygens had received the complimentary copy sent to him by Newton before the end of 1687.[3] In the summer of 1689 Huygens himself visited England for the second time, his brother Constantin having gone over in the service of William III. On 22 June he attended a meeting of the Royal Society where he was introduced to Newton. Several other meetings took place between them, notably on 10 June, when Huygens waited on the King with Newton to press the latter's claim to the headship of King's College.[4] It was probably in August that they discussed the mathematical representation of motion in a resisting medium, their conversation resulting in some pages of notes by Newton explaining his position, with marginal comments by Huygens.[5]

In these notes Newton gives a detailed exposition of the fundamentals of his theory of resistance. He thought that resistance could be treated as what would now be called a vector quantity, that is to say that the 'absolute' resistance experienced by a body moving along a straight line could be treated mathematically as the product of two 'oblique' resistances operating along the other two sides of a right-angled triangle along the hypotenuse of which the body moved. (This, however, Huygens would not accept.) The main point at issue between them, however, was that Newton

[1] 'Possemus etiam in unum componere resistentiam absolutam . . . & respectivam . . . uti certe revera concurrunt in natura, sed prolixitas hic vitanda est. . . . Omnia autem respondent nostrae Analysi infinitorum hoc est calculo summarum et differentiarum (cujus elementa quaedam in his actis dedimus) communibus quoad licuit verbis hic expressis.'

[2] 24 June 1687. *Œuvres*, vol. IX, pp. 167–9.

[3] *Ibid.* pp. 190–1, 267, 305.

[4] *Ibid.* p. 333.

[5] *Ibid.* pp. 321 *et seq.* First printed by Johann Groening in *Historia Cycloidis* (1701).

asserted, while Huygens denied, that the trajectory of a projectile in a resisting medium has a vertical asymptote. Newton showed that, according to his principles, if two bodies descend from *R* the one along *RS* and the other vertically along *RT*, reaching the

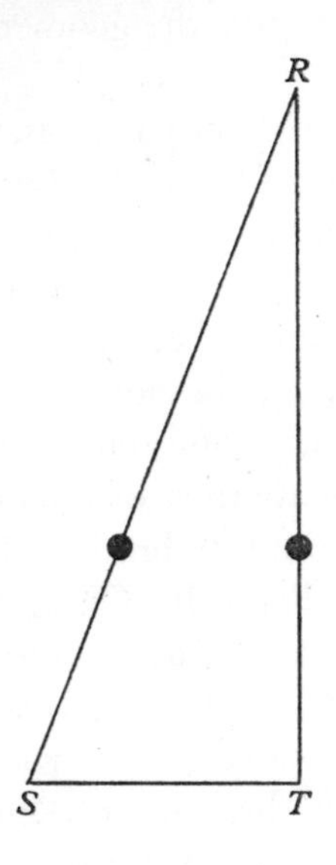

Figure 9

horizontal line *TS* at the same instant, the oblique resistance in the direction *RT* to the body descending along *RS* is to the direct resistance along *RT* opposed to the other body as RS^{n-1} to RT^{n-1}, if the medium resists as the nth power of the velocity, or as *RS* to *RT* if the resistance is proportional to the square of the velocity. Suppose that *RS* is not a straight line but a very small portion of the descending branch of a trajectory; then before the falling projectile can attain its terminal velocity it must descend perpendicularly, since an oblique perpendicular resistance is greater than a direct one.

In other words, as in an infinite time it tends to reach a (maximum) terminal velocity, so its descent tends to become indistinguishable from the perpendicular. On this passage Huygens commented:

On the contrary I place such velocity downwards because it experiences a resistance equal to gravity so that consequently the downwards acceleration is zero, and because retardation occurs in lateral motion, it follows that there is retardation in oblique or absolute motion.

Newton's argument in the *Principia* and his likening of the trajectory to a hyperbola had failed to convince Huygens that the ballistic curve has a vertical asymptote, and he was not moved by these later reasonings from first principles which he seems to have understood no more than he had Newton's early papers on light, though he was quite correct in stating that resistance cannot always be treated as a simple vector quantity. Both Leibniz and Bernoulli, however, agreed with Newton that the trajectory is asymptotic to a perpendicular line.

When, however, in the 'Addition' to the *Discours de la Pesanteur* (published along with his *Traité de la Lumière* in 1690) Huygens made known the results of his calculations and experiments in 1668-9, he did not refer to the points of disagreement between himself and Newton.[1] The *Discours* had certainly been in preparation for some time since there is mention of a 'Tractaet van de Oorsaeck der Swaerte' in a letter to the directors of the Dutch East India Company of April 1688, thus associating it with the period when Huygens was studying the *Principia*.[2] After some remarks on Newton's theory of the shape of the earth and his general theory of universal gravitation, Huygens declared that he had read with pleasure Newton's treatment of resisted motion, having formerly applied himself to the same investigations.

Et puisque cette matière appartient en partie à celle de la Pesanteur, je crois pouvoir raporter icy ce que j'en découvris alors. Ce que je ne feray pourtant qu'en abregé & sans y joindre les demonstrations, ayant negligé de les achever, parce que cette speculation ne m'a pas semblé assez utile, ni de consequence, à proportion de la difficulté qui s'y rencontre.

He explained how his first calculations had been rendered abortive by his discovery of the square law of resistance, which the corpuscular structure of matter made sufficiently probable. This discovery had complicated his calculations, in which he had used the logarithmic curve as his means of analysis, but he had been able to attain certain results, which he proceeds to describe. In

[1] *Traité de la Lumière . . . avec un Discours de la Cause de la Pesanteur* (Leyden, 1690). The 'Addition' extends from pp. 152-80 (Letter to Fatio de Duillier, Feb. 1690, *Œuvres*, vol. IX, p. 358).

[2] *Ibid.* p. 276.

addition to the relations mentioned in the account of Huygens' work in a previous chapter, he here proved that his construction of the trajectory when the resistance of the medium was supposed proportional to the velocity produced the same curve as that outlined by Newton.

Huygens wrote a letter describing his latest publication to Leibniz, which began a lengthy correspondence between them on resisted motion and the progress of mathematics in general,[1] for Leibniz during two years of travel and preoccupation with political affairs had been somewhat cut off from the news of the scientific world. He was prepared to believe that Newton's theorems agreed with his own views announced in the *Schediasma*, and that what he had called respective resistance was the same as Newton's and Huygens' medium resisting as the square of the velocity, since it was proportional to velocity and time jointly. He therefore asserted that Huygens' results were in no way except the expression different from his own. But the latter failed to see that Leibniz's hypothesis was the same as that of Newton and himself, in fact he thought that the whole of his article was obscure and unintelligible.

During this exchange of letters Huygens' attitude to Leibniz was more than a little cold, because he suspected that Leibniz privately esteemed the differential calculus far above his own methods, and was not above suggesting that Huygens had borrowed from him. Leibniz, on the other hand, anxious to enrol Huygens along with other continental mathematicians against Newton and the English school (though this was long before the great priority dispute broke out), in spite of Huygens' membership of the Royal Society and friendship with many English scientists, went more than half way towards conciliation, denying Huygens' suspicions and even correcting some of his propositions in deference to Huygens' criticisms, 'corrections' which he had to retract at once because he found that they were themselves wrong. He laboured hard and finally with success to persuade Huygens, who was now of an age to be conservative, of the value of differential methods and transcendental equations.

It must be admitted that Leibniz was patient, for the whole

[1] Feb. 1690 (*ibid.* p. 366).

of the mathematical discussion between himself and Huygens was complicated by the alien characters of their methods; whereas Huygens made use of a development of conventional geometry with no special notation by which he was virtually able to differentiate and integrate from first principles, Leibniz wrote in the notation of the calculus which Huygens did not understand, and which to him was undoubtedly obscure. His careless slips in calculation provoked Huygens, his rather loose claims in private correspondence to the published discoveries of others irritated Huygens as they did Newton, and he was also blamed for an unsympathetic review of the *Traité de la Lumière* in the *Acta Eruditorum*. There was no quarrel; gradually under the instruction of its inventor Huygens conceded the powers of the new analysis, while it was important for Leibniz to receive the approbation of his old master in mathematics, now esteemed as the doyen of European science.

Even so, Huygens could not see that Leibniz' resistance proportional to the velocity and the elements of time was the same as his own resistance proportional to the square of the velocity. He regarded resistance as a physical fact akin to gravity, a pressure upon the moving body, which could be thought of independently of time. It became clear to him that whatever the differences in mathematics between Newton, Leibniz, and himself, he shared with Newton a physical interpretation quite different from that of Leibniz. At this time (February 1691) he was revising his old calculations and comparing them with the ones that others had made since, and he informed Leibniz:

Je trouve qu'une partie de notre dispute vient de ce que vous prenez le mot de resistence dans une autre signification que moy et M. Newton; car vous appellez resistence la velocité perdue, . . . et en consequence de cela pour comparer des resistences differentes, vous voulez que la consideration des elemens du temps entre en compte, et 'qu' à parler exactement, on ne doit pas dire que les resistences sont en raison des velocitez ni en raison des quarrez des velocitez'. En quoy il est evident que vous prenez l'effet de la resistence pour la resistence mesme. Mais à M. Newton et à moy la resistence est la pression du milieu contre la surface d'un corps, comme par exemple, quand on tient dans la main une feuille de carton, et qu'on l'agite à travers l'air, on sent une presaion qui si peut comparer a celle d'un poids, et qui devient quatre fois plus grande lorsqu'on remue cette feuille deux fois plus vite qu'auparavant, ainsi que j'ai trouvé autrefois à Paris par des experiences fort exactes. Vous voiez, Monsieur, qu'il n'y a que la differente vitesse dont depend cette pression, sans considerer des parties égales ni inégales des temps. Et c'est sans doute la veritable et la plus naturelle notion de la resistence.

He softened his critique with a reference to Leibniz's manifold activities as historian, philosopher, and politician which prevented him from concentrating on mathematics; he knew that Leibniz did not lack ability to unravel the problems of resistance, but time to bring more clarity and precision to the discoveries he had made. He suggested that Leibniz should carry his researches further than he had in the *Schediasma*, when he would find that the ballistic curve was not so simple as he supposed in that article, principally because if the resistance was as the square of the velocity it could not be resolved into components (vectors).[1]

In revising his previous work on ballistics Huygens did not break any new ground or come across any fresh results fit for publication, but he was able to satisfy himself that it agreed with the theorems of Newton and Liebniz and to correct several numerical slips they had allowed to pass. Expressed in modern form his basic relation was

$$t=\frac{V}{2g}\log_e\left(\frac{V+v}{V-v}\right)$$

where v is the velocity of a body falling freely through a medium resisting as the square of the velocity after the time t, and V its terminal velocity.[2] Consequently he found the distance fallen s in the time t to be

$$s=2gt^2\frac{\log_e\left(\frac{V^2}{V^2-v^2}\right)}{\left[\log_e\left(\frac{V+v}{V-v}\right)\right]^2},$$

which substituting for t becomes

$$s=\frac{V^2}{2g}\log_e\left(\frac{V^2-v^2}{V^2}\right),$$

[1] Huygens to Leibniz, 23 Feb. 1691 (*Œuvres*, vol. x, pp. 17 *et seq.*). See also vol. xxi, pp. 423–5.

[2] 'De descensu corporum gravium et ascensu per aerem aut materiam aliam quae resistit motui in ratione duplicata celeritatum, ut revera contingit. Olim inventa clarius hic explicare volui...' (*ibid.* vol. x. pp. 23 *et seq.*).

and so with this equation was able to correct the first corollary of Prop. IV of Book II of the *Principia*, where Newton had made a mistake in naming the points of his figure, as it was corrected by Newton himself in the second edition. He noticed carefully a number of similar rather trivial mistakes in Leibniz's calculations sent him during their correspondence, in which he had expressed analytically his theorems given in the *Schediasma*, 'et tamen saepius jam calculum suum correxerat'.[1] His other equations, being identical with those discussed in a previous chapter, need not be repeated here.

Leibniz also in the spring of 1691 was going over his theory of resistance, mentioning to Huygens his intention of publishing something in the *Acta* on the three studies of resistance made by Huygens, Newton and himself.[2] This took the form of an 'Additio ad Schediasma de Medii Resistentia publicatum in Actis mensis Febr. 1689'.[3] Here Leibniz declared that after the publication of his first article, reading the discussions of the same problem by Newton and Huygens, he had found that they limited their treatment to respective resistance, 'qualem scilicet sentit corpus in liquido tenacitate notabile carente, velut in aere,' not considering the absolute resistance arising from the viscosity of the medium or the friction of its particles which he had spoken of first. So far as respective resistance was concerned, their results agreed, for they had taken the resistance to be proportional to the square of the velocity,[4] while he had written that it was 'in ratione composita velocitatum & elementorum spatii', which was the same thing, because over equal time intervals the spaces are proportional to the velocity. Furthermore, though they had each expressed the relation between time and velocity in the free fall of a heavy body through a resisting medium in a different way, by the hyperbola (Newton), by an infinite series (Huygens) and by a logarithmic curve (Leibniz), these various methods hid an identical mathematical relation. As for Huygens' warning that resistance could not be treated as a vector, he admitted that it was true, except in the case of a particular hypothesis, such as the fall of a weight

[1] *Œuvres*, vol. x, p. 38.
[2] *Ibid.* p. 9.
[3] *Acta Eruditorum*, April 1691, p. 177. Published over the initials O. V. E. Reprinted, Gerhardt, *op. cit.* p. 143.
[4] This was only partly true of Newton.

in the hold of a ship, where the medium itself is in motion.[1]

On the whole the mathematical labours involved in this correspondence had afforded little promise that Leibniz's calculus could solve the mystery of the ballistic curve. He had found out nothing that Newton had not already published, and that Huygens had not discovered with the aid of conventional mathematics, while his oversights and retractions did not inspire confidence. In fact it was only the solution of the catenary problem by Leibniz and Bernoulli in the autumn of 1691 that really convinced Huygens that the new mathematics offered advantages over the old and could arrive at the same precise results.[2] But by the time of the next trial of skill the ascendancy of the differential method was marked.[3]

In 1699 Fatio de Duillier published an attempt at the curve of least resistance, in which he required five and a half pages for his lengthy solution—and incidentally began the priority dispute.[4] A copy of this was sent to the Marquis de l'Hôpital, who thought it easier to work out a solution than to wade through Fatio's laborious calculations.[5] Johann Bernoulli, who published a third analysis of the problem leading to an identical equation three months after the Marquis' solution was printed, was even more outspoken than he had been in criticism of Fatio, whom, in a manner soon to become odiously familiar, he accused of incompetence as a mathematician.[6] To find the curve of the solid of least resistance, he declared, was not a problem more difficult than that of the brachystochrone, nor was it necessary to employ second differences, as Fatio had done, since the differential

[1] Leibniz's special pleading in defence of the *Schediasma* is scarcely convincing.

[2] Huygens to Leibniz, 1 Sept. 1691: 'Je consideray ensuite pourquoy plusieurs de vos découvertes m'estoient echappées et je jugeay que ce devoit estre un effet de vostre nouvelle façon de calculer, qui vous offre, à ce qu'il semble, des veritez, que vous n'avez pas mesme cherchées, car je me souviens que dans vos lettres precedentes vous m'aviez dit, en parlant de ce que vous aviez touvé touchant la *Catenaria*, que le calcul vous offroit cela comme de soymesme, ce qui certainement est fort beau' (*Œuvres*, vol. X, p. 129).

[3] Montucla, *Histoire des Mathematiques* (1758), pp. 460 *et seq.*

[4] *Nicolae Fatii Duillierii R. S. S. Lineae Brevissimi Descensus investigatio Geometrica duplex. Cui addita est investigatio Geometrica solido rotundi in quod minima fiat resistentia* (London, 1699). Bound with his *Fruit Walls improved.*

[5] It was published in the *Acta* (August) and the *Mémoires de l'Académie Royale des Sciences* for the same year.

[6] This was not then the opinion of Huygens and Newton, nor later of Cantor (*Vorlesungen*, vol. III, p. 248).

equation $y\ (dy)^3\ dx = a\ (ds)^4$ was appropriate to it.[1] Fatio, however, was not able to rival their results by fluxional methods until 1713.[2]

In the early years of the eighteenth century the ballistical questions which had exercised the ingenuity of mathematicians yielded very quickly to the new methods of analysis, preparing the way for the foundation of modern ballistics in the second half of the century. The first stage of this process was a long series of articles by Pierre Varignon in the *Mémoires* of the Parisian Academy from 1704 to 1711, in the first of which he gave the general equation for the motion of a body acted upon by two forces (*impressions*) simultaneously: $v\,dy = z\,dx$—'laquelle deviendra celle de telle hypothèse qu'on voudra, si l'on y substitue en x et en y & en constantes les valeurs des vitesses v & z qui conviennent à cette hypothèse'. For instance, he says, in the case of a body projected with a uniform velocity $z = \sqrt{a}$, and descending under gravity at a velocity whose square is proportional to the distance fallen, so that $v = \sqrt{x}$, substituting in the formula we get $\frac{dy}{dx} = \sqrt{\frac{a}{x}}$ of which the integral is $4ax = y^2$, the ordinary equation for the parabola.[3]

In his next paper he summarised by means of the calculus the propositions already put forward by Newton, Leibniz and Huygens on the rectilinear movements of bodies in resisting mediums,[4] and in 1708 published a demonstration of Huygens' constructions of the trajectory where the resistance is proportional to the velocity given in the *Discours de la Pesanteur*, showing at the same time the conformity of all results previously published and his own. Next year he contributed a paper on the curve that would be produced on the assumption that Galileo's law of acceleration was exactly true and that the velocity of projection only was sensibly reduced

[1] *Acta Eruditorum*, Nov. 1699, pp. 513 *et seq.*: *Excerpta ex literis Dn. Joh. Bernoullii Groningae 7 Augustis 1699 datis.*

[2] Bernoulli published a second article noting the agreement of his solution with that of the Marquis in May 1700. Fatio replied in the *Acta* (March 1701), but did not complete his first solution till 1713.

[3] 'Manière de discerner les vitesses des corps mus en lignes courbes: de trouver la nature ou l'équation de quelque courbe que ce soit engendré par le concours de deux mouvemens connus' (*Mémoires*, 1704), pp. 286 *et seq.*

[4] 'Des mouvemens faits dans des milieux qui leur resistent en raison quelconque' (*Mémoires*, 1707), pp. 382 *et seq.*

by the resistance of the medium, continuing into 1710 and 1711 his investigations of the variations in velocity caused by the resistance of the medium.

This treatise on resistance, for such the several memoirs amounting to some hundreds of pages and utilising the most advanced mathematics of the age really formed, marks the end of the transitional stage of the application of mathematics to ballistics. A new stage began with the publication of Johann Bernoulli's general solution of the ballistic problem in 1719. Possibly he would never have discovered this theorem but for the bitter dispute between himself and John Keill, Savilian Professor of Astronomy at Oxford, over the rival merits of Newton and Leibniz, which soon degenerated into a low squabble about their own mathematical powers. The *Acta Eruditorum* had been freely used by Leibniz and Bernoulli to make public their own version of what had happened in 1676 and since in the development of mathematical analysis, while from 1705 Keill had been using the *Philosophical Transactions* and later the *Journal Littéraire de la Haye* to present his defence of Newton mingled with counter-charges against his critics. The controversy had really become an entirely personal one between Keill and Bernoulli after the anonymous publication of the latter's 'Epistola pro Eminente Mathematico D. Johanne Bernoullio contra quendam ex Anglia antagonistam scripta' in 1716, of which, when he read it in May 1717, Keill wrote to Newton from Oxford that he believed 'there was never such a piece for falsehood, malice, envy and ill-nature published by a mathematician before', promising 'to set about answering him as he deserves'.[1] This answer took the form of an open letter to Bernoulli, written with Newton's assistance and approval, and published in the *Journal Littéraire* for 1719, which was not actually printed until 1720.[2]

Thereafter, apparently, Bernoulli wrote a letter to Newton which gave Keill reason to declare 'he is sensible he has burnt his fingers and would be glad to get off, but if he be in earnest he ought to beg your pardon publicly'.[3] The Oxford professor,

[1] Cambridge University Library, MSS. Add. 4007 (transcripts), dated 17 May 1717.

[2] A fair copy of the letter, incorporating the notes and additions written by Newton on a separate sheet (which were not in the direction of moderation) is in the Portsmouth Collection (Add. 3968, no. 23). [3] Transcript in Add. 4007, dated 23 May 1718, to Newton.

however, had already despatched a challenge to Bernoulli which reached him early in 1718. It is quite clear that Keill, in defying him to 'Find the curve which a projectile describes through the air on the simplest hypothesis of uniform gravity and density in the medium, the resistance varying as the square of the velocity', was guided not so much by a wish to advance mathematics and ballistics as by a desire to humiliate Bernoulli by means of a problem which he supposed to be insoluble, and Bernoulli was quite justified in viewing the challenge in this light. Many such challenges had been issued in the past, even recently by himself, but in every case they had been of a general character, aimed at no particular person, and the challenger had first discovered the solution for himself.

It was natural therefore that Bernoulli, when he published his solution of Keill's problem and an account of his conduct, should dwell at greater length upon his triumph over the English mathematicians than upon his very considerable achievement in mathematics.[1] His character was not an amiable one, and his part in the calculus controversy had been neither judicious nor unblemished, but in this instance he had some excuse for castigating the 'indecent provocation' of a certain Scotsman whose manners had already made him detestable. Keill's challenge could have no bearing upon the question originally at issue, whether Newton or Leibniz was the first inventor of the calculus, nor could it prove Keill to be a more accomplished mathematician than Bernoulli, since he failed to produce an answer to his own problem. Bernoulli might well wonder that Newton, whose reputation 'all who are true judges of fine things admire', should allow such a champion to take his part. As for the corrections to the *Principia* which he had published, Bernoulli wrote, Newton himself had acknowledged them in the second edition of the work, and in any case was it not proper to substitute truth for error? But Keill had been absurdly annoyed as though the whole of the island race had been insulted because a foreigner had taken upon himself to put a Briton right. His attempt to make the critic

[1] 'Responsio ad nonneminis provocationem, ejusque solutio quaestionis ipsi ab eodem propositae, de invenienda linea curva quam describit projectile in medio resistente' (*Acta Eruditorum*, May 1719), pp. 216 *et seq*.

look ridiculous by proposing to him a problem the inverse of which Newton himself had found impossible of precise solution had failed because it had happened to him to discover what had escaped even the lynx-eyes of Newton. Having received Keill's challenge in February 1718 he had shortly after found a solution of an even more general character, and he had in turn through a third party required Keill to 'Construct the curve (the quadrature being granted) which a heavy body tending to fall perpendicularly to the horizon describes in a medium of uniform density; supposing the resistance to be as any power of the velocity of which the exponent is 2n'. This Keill failed to do, though after the final date had passed a solution had been submitted by Brook Taylor (an English mathematician justly better known than Keill).

After his history of the problem presented to him Bernoulli related his own result, wisely refraining from publishing his method lest it be used against him, though he privately communicated it to De Montmort and Varignon. Let z be an indeterminate number; construct the area $\int (a^2+z^2)^{n-\frac{1}{2}}\, dz$ and let this be represented by Z; then, said Bernoulli, if x, y, are the ordinates of the required curve,

$$x=\int z\,dz\,Z^{-\frac{1}{n}}; \quad y=\int a\,dz\,Z^{-\frac{1}{n}}$$

On this general solution he gave corollaries. In Keill's problem $n=1$, and the curve (as Newton had said) depends upon the quadrature of the hyperbola. If n is an odd number greater than unity the curve is a transcendant of the first degree. If it is a negative odd number the curve depends upon the quadrature of the circle. (This again Newton had proved.) If $n=\frac{1}{2}$, the curve is a logarithmic, for which he offers a construction neater than those of Huygens and Newton.

The subsequent history of mathematical ballistics relates the many attempts to integrate approximately the fundamental differential equation given by Johann Bernoulli, while including in the calculation the various physical complexities outside his simple hypothesis which have been found necessary in the course of the experimental study and technical development of gunnery. By the use of the laws of motion, the principles of mechanics and the subtleties of mathematics the greatest mathematician of the seventeenth century had produced the earliest and partial solution of the chief theoretical problem of ballistics. To rewrite this solution in usable terms and to apply it in practice was the work of a period in which the art of calculation and machine-making had advanced far beyond the rudimentary practices of the age of Newton, and of a society whose purposes and ambitions made it very different from that in which he lived.

CHAPTER VII
CONCLUSIONS

The beginnings of a science of ballistics, as of so much else, are to be found in Aristotle, who introduced the theory of the motion of projectiles into his explanation of the universe because it seemed to him a distinct and peculiar species of motion; and it was important for his physics to show that, in this as in other species, there is an assignable relationship between cause and effect. Aristotle's doctrine was rightly regarded in later times as a piece of reasoning in philosophy, a fragment in the mosaic of learning; as such and no more it was transcribed and commented upon, until in the fourteenth century it was subjected to the 'experimental' judgements of the impetus theory, followed in the early sixteenth by the second 'experimental' application to ordnance. Only then, after two thousand years, was the connection between the philosopher's theory of projectiles and the soldier's use of them appreciated, in a way characteristic of the new course of European thought from which science was born.

Then there took place in the sixteenth and seventeenth centuries the triumphant progress of mechanical science, only rivalled by the advancement of pure mathematics which helped to make it possible, and culminating in the *Principia*. In this progress ballistics played a minor but not insignificant part. This limited branch of science served as a mirror reflecting much of the most advanced scientific thought. Philosophers from Galileo to Newton—whose active lives embrace the whole of the seventeenth century—used the problems of ballistics as a gymnasium in which to develop their powers for larger and more important researches. Here were illustrations for the two primary definitions in science, inertia and acceleration; here Galileo exemplified the scientific method of abstraction; here also were profitable exercises in the combination of experiment and hypothesis and in the application of mathematics to natural philosophy. Indeed it may be said that modern science, arising from the impetus philosophy of the fourteenth century, owes its origin to the most obvious fallacy in Aristotle's physics, his theory of projectiles. But with the opening of the eighteenth century the usefulness of ballistics to general science ended, for Newton had explored to its limit Galileo's method of treating celestial and terrestial mechanics as parallel paths of progress.

It may happen by the process of extension and exhaustion that the direct heir to the great ideas of the past is a trivial problem of detailed application, while the main artery of living knowledge has branched out; thus the development of impetus mechanics from Buridan to Leonardo, or the formulation of dynamics by Galileo in the theory of projectiles are great events in the history of science, compared with which the publication of Bernoulli's paper in 1719 is insignificant. Ballistics soon became what it had not been before, an isolated compartment of applied physics and mathematics, intimately related to technology. The labour of removing the immense mathematical difficulties in the way of practicable solutions of its problems, and the accumulation of experimental data, remained to tax the abilities of leading mathematicians like Euler, but the interest of such work was strictly limited. Moderate success was only achieved towards the end of the century, and its application to the art of war awaited the engineering revolution.

The reasons why the mechanical sciences pushed ahead more rapidly than other branches are not really obscure. Only in mechanics and mathematics was it possible to build securely on the foundations laid in antiquity and the Middle Ages, and for original minds to take up the task without first clearing away the system of errors which was elsewhere their only legacy. Archimedes, the founder of statics, was accordingly chosen as the ancient pattern of the natural philosopher, for the rigidity of his demonstrations derived from mathematical, experimental methods. To the end of the seventeenth century botany and biology were still struggling with the preliminary problem of classification; progress in medicine was fortuitous while physiology and anatomy remained in a rudimentary condition; in chemistry, until the need to abandon medieval ideas was apparent, analysis could not begin;

physics consisted of disconnected fragments of experimental knowledge. Only in mechanics could first principles be deduced from relatively simple observations with the aid of mathematical considerations already understood, and it is no discredit to Galileo or Newton that their great discoveries could not have been duplicated in any other part of science. In short the materials were prepared for a rapid advance in mechanical science, by Stevin in statics and by Galileo in dynamics, whereas in other sciences a long gradual effort of preparation for a crux was still to come.

Thus when conditions purely internal to science and the still powerful tendency to think of cosmology as the core of science favoured the creation of dynamics as the first fully 'modern' branch of science, it is unnecessary to look further afield for material motives. Men were led to discoveries in mechanics less by their practical usefulness than by the logic of historical development and the relative ease of success in that part of science at that stage in its evolution. The theory of projectiles was pre-eminently a traditional crux in philosophy, and whatever else may be said of Galileo, it should never be forgotten that always his driving force was a passionate desire to correct the errors of the conventional doctrine. From that desire Galileo's interest in ballistics sprang, and his triumph was perhaps more satisfying to him than anything else he did.

But for that, and the progress of mathematics in the hands of Newton and Leibniz, a useful science of ballistics could never have been created. The progress of pure mathematics was itself illustrated by its application to mechanics, on the grand scale in the *Principia*, in a lesser way by mastering problems like the cycloid, the catenary and the brachystochrone. The challenges offered to each other by seventeenth-century mathematicians were almost all put into the concrete form, and the problems of mechanics, astronomy included, offered the most fertile field for the use of advanced methods. In directing their attention to them mathematicians naturally selected those suggested by the background and experience of their age. The theory of projectiles was an obvious subject for mathematical exercise as it was brought to mind by the most striking instance of the power commanded by man. Things in which no one is interested are not likely to become instruments for demonstrating the powers of new methods; and to that extent the type of practical problem which the mathematician took up, including the problems of ballistics, was conditioned by the experience of an age in which projectiles and navigation were prominent, whereas heat-engines (for instance) were not. But the usefulness of the solution is a secondary consideration; we may say of a mathematician speculating on the catenary that without his commonplace experience of suspended ropes his attention would not have been drawn to the possibility of analysing such curves mathematically, yet we are not necessarily to believe that the way in which ropes hang is of importance to society. Assuming that the phenomenon is within the general experience, and therefore a potential subject for scientific treatment, the moment when it will be treated is determined only by the development of science to a competent level.

The truth of this is sufficiently illustrated by the history of ballistics. The first step towards modern mathematical science, the idea of impetus, was taken before there was any considerable change in the art of war, in the form of a philosophical revolution within the universities. When artillery did emerge in the late fifteenth century as an impressive, notorious phenomenon, the prevailing philosophy was simply adopted by technical writers who fitted facts to theory, rather than theory to facts. A recent attempt by E. J. Walter[1] to make the refoundation of dynamics dependent upon the invention of artillery is thus controverted by the clear evidence of chronology. Military writers were not original thinkers to whom Galileo was deeply indebted; on the contrary they themselves were plagiarists from a long philosophic tradition. The practice of artillery contributed nothing to seventeenth-century science but the convenient illustration of dynamical principles, while the practical men, having borrowed one theory, clung to it long after the thought from which it was derived was dead. Men had been pondering the motion of projectiles for thousands of years; approximately correct ways of thinking about it were only possible when correct mechanical principles and an appropriate mathematical method had been attained. It is true that in the

[1] 'Warum gab es im Altertum Keine Dynamik?' *Archives Internationales d'Histoire des Sciences*, No. 3 (April 1948), p. 365.

sixteenth and seventeenth centuries the theory of projectiles offered the most useful and familiar approach to these principles and methods, but it is debatable at least whether that approach was confirmed by thoughts of material interest.

Those who maintain the contrary have principally been writers of what may be termed the sociological history of science, in itself an admirable part of the subject, but one not without its pitfalls. While historians of mathematics or biology have worked out the logical development of their studies, showing how ideas have succeeded each other in order of complexity, and seeking to explain by an internal view of science why at one period a profitable suggestion has been ignored or a clumsy hypothesis prolonged, social historians have attempted to define changing attitudes to science and to use the picture of a society at a given epoch to account for the relative popularity of one branch and the stagnation of another. The two methods are complementary, and in extremes exclusive, for the former isolates the scientist from the world in which as a man he lives, the latter places him at the mercy of his environment, and makes a great discovery no more than the response of a quick mind to the most pressing need of the moment. The danger of this last point of view is that it overlooks the real difference between the apprehension and judgement of a given set of facts by the scientific mind considering them for its own purposes and that of an ordinary manufacturer, craftsman or citizen; further, it underestimates the vast changes in society which have not been sought, but have resulted accidentally as a by-product of increasing scientific knowledge.

In seventeenth-century ballistics we have seen already that the mathematician moved in his own order of concepts. Neither soldiers, nor craftsmen, nor governments were very much interested in the scientifically important problems, and though technological applications may at once suggest themselves to us, they were not then desired nor practicable. At the time of their composition Newton's propositions were as irrelevant to the technical practice of the age as Maxwell's electromagnetic waves; and the practical applications of the one and the other were equally unforeseen. The history of industry frequently reveals the survival of techniques scientifically obsolete and inefficient; that of science repeatedly shows the refinement of theories already surpassing technological development. How far can we reconcile these contrasting attitudes? Only by assuming that advance in science never happens unless accompanied by an accelerating industrial progress, an assumption which reduces to the belief that no one ever strives for knowledge which is not of material use. Thus when Hessen[1] wrote that the scheme of seventeenth-century physics 'was mainly determined by the economic and social tasks which the rising bourgeoisie raised to the forefront' he was guilty of a gross anachronism, since the level of physics which Newton approaches in the *Principia* was only reached by an original intellectual process, not directly from the empirical industry of his day; only after engineering had transformed transport, manuture and war in the early nineteenth century were their problems raised to the order of refinement for which Newtonian mechanics were appropriate.

The social historians have doubtless been influenced by the seventeenth-century literature of apology for the new philosophy, in which, from the time of Bacon onwards, utility is put forward as a principal reason for the study of natural science. Long life, prosperity and wealth were expected from the command over nature which science could confer, and it is only by remembering that this optimism was not unrelated to a belief in the possibility of discovering the secrets of the philosopher's stone, the transmutation of metals, and perpetual motion, that it can be appreciated as nothing more novel than the public avowal of the esoteric ambitions of the medieval alchemist. Few of the public apologists for science were themselves scientists. The question at issue is not whether there was a feeling in society that philosophers should study 'things of use' instead of weighing the air and magnifying the flea, but whether the philosopher himself yielded to such an opinion. The evidence seems to show that he did not; he clung to his old name and status and resisted the attempt to make natural philosophy a sort of superior technology restricted to the problems of cider-making and navigation. Theoretical studies held the field —the scientist could be a Gimcrack if he chose, so that by the end of the century the interest aroused by publicists like Samuel

[1] *Science at the Cross Roads* (London, 1931).

Hartlib had waned because the promised miracles had not occurred. The apologists of science in their enthusiasm unconsciously destroyed the very medieval ideas of nature and 'natural magick' to which they and not the new type of mathematical philosopher were still attached, and which alone were the foundation for their hopes of a short cut to material progress.

The real opportunity for the mutual interaction of science and techniques occurs in the field of invention, but the professional adaptor of scientific principles to commercial ends is an entirely modern figure. If we treat seventeenth-century inventiveness as a real indication of the needs and interests of the century, the contrast with science becomes plain. Military inventions were not numerous; commercial activities such as transport and building attracted more attention. Inventions tending to improve the ballistics of ordnance will be sought for in vain; startling and horrific devices, the perennial war-chariot, submarine vessels and automatic weapons were inventions repeatedly attempted without success. There is simply no relationship at all between the theory of ballistics, which was an integral part of the revolution in dynamics, and the implausible proposals for destruction or defence which were the natural products of a warlike but unmechanical age. Where the soldier felt a need for new methods of making armaments more durable and powerful, fertile inventors offered largely useless extravaganzas, and the scientist nothing; where he could make effective use of his own procedures in the directing of shots to a target, the scientist made available a store of knowledge which the soldier did not want and could not use. External ballistics was utterly useless for the ordnance of seventeenth-century armies and navies, and only a complete misconception of the state of science and technology can induce the belief that the theorems of Huygens, Newton, Leibniz and Bernoulli were of any practical value. Just as Newton's theory of gravitation was more than an account of the fall of an apple, so mathematics had sublimated ballistics out of the sphere of the Board of Ordnance. If we imagine that scientists were furthering the cause of technological evolution in this respect, we attribute to them a prescience of the industrial revolution which altered the techniques of gun-founding and gunnery, more than a century later.

The scientific revolution sprang from quite different sources, and followed an altogether separate channel, from the new awareness of the importance of technology, in spite of the efforts of publicists to bring the two together, and the colour given to their assertions by some genuine scientists. The fact of a common curiosity in why and how things happen is no proof of a common purpose to turn all explanations to utilitarian ends. Even in ballistics, curiosity and tradition being important in directing attention to what was, after all, the most impressive and potent of all man's works, the links between scientific researches and contemporary estimations of technical needs were but slight. Ballistics was a science necessarily created in the course of the scientific revolution. It flourished or languished as the aspects of physical science and mathematical technique upon which it depended prospered or were neglected, offered easy results or opposed difficult obstructions. As an element in continuous historical tradition of science the theory of projectiles found a place in the two greatest treatises of physical science, and as such it should be considered by the historian: as naturally offered by science in the course of its evolution, not wrenched from it by the strong hand of economic necessity.

APPENDIX

Artillery Tables

I. *English*: *c.* 1590

This was drawn up by John Sheriffe and is to be found in S.P. 12/242, nos. 64, 65. It is possible that it represents a rather antiquated practice.

'The Brevity and the secret of the Art of great ordnance necessary for all generals for their present memory.'

		Height of Piece[1]	Weight of Piece	Weight of Shot	Weight of Powder	Point-blank Range	Extreme Range
These pieces be most serviceable for battery being within 80 paces of their mark which is the chief of their force.	Cannon Royal	$8\frac{1}{2}$ in.	7000 lb.	66 lb.	30 lb.	320 paces	1930 paces
	Cannon	8	6000	60	27	340	2000
	Cannon Serpentine	$7\frac{1}{2}$	5500	$53\frac{1}{3}$	25	400	2000
	Bastard Cannon	7	4500	$41\frac{1}{4}$	20	360	1800
	Demi-cannon	$6\frac{1}{2}$	4000	$30\frac{1}{4}$	18	340	1700
	Cannon perrier	6	3000	$24\frac{1}{4}$[2]	14	320	1600
These pieces be good and serviceable to be mixed with the above Ordce. for battery to pierce being crossed with the rest as also fit for castles forts and walls to be planted for defence.	Culverin	$5\frac{1}{2}$	4500	$17\frac{1}{3}$	12	400	2500
	Basiliko	5	4000	$15\frac{1}{4}$	10	none stated	—
	Demi-culverin	$4\frac{1}{2}$	3400	$9\frac{1}{3}$	8	400	2500
	Bastard culverin	4	3000	7	$6\frac{1}{4}$	360	1800
These pieces are good and serviceable for the field and most ready for defence.	Saker	$3\frac{1}{2}$	1400	$5\frac{1}{3}$	$5\frac{1}{3}$	340	1700
	Minion	$3\frac{1}{4}$	1000	4	4	320	1600
	Falcon of $2\frac{1}{3}$	$2\frac{1}{3}$	800	3	3	300	1500
	Falconet	2	500	$1\frac{1}{4}$	$1\frac{1}{4}$	280	1800

II. *Spanish: c.* 1603

Points of the Quadrant[1]	Point-blank	1	2	3	4	5	6
	Ranges (paces)						
Cannon of 50 lb.	1000	2500	3900	4000	4486	4550	4660
Cannon of 45 ,,	950	2400	3800	3947	4464	4620	4700
Cannon of 40 ,,	900	2220	3700	4316	4490	4780	4792
Cannon of 35 ,,	850	2040	3570	4284	4613	4766	4834
Cannon of 30 ,,	800	1866	3421	4227	4636	4814	4900
Demi-cannon of 25 lb.	750	1600	3200	3800	4434	4800	5600
Demi-cannon of 20 ,,	700	1540	3080	3620	4090	4526	5389
Demi-cannon of 12 ,,	600	1280	2560	3400	3980	4246	4380
Third-cannon of 12 ,,	500	1033	2066	2581	3064	3300	3300
Culverin of 22 lb.	800	1737	3466	5548	6469	7120	7355
Culverin of 20 ,,	720	1560	3150	5000	5990	6587	7200
Culverin of 18 ,,	700	1487	2974	4516	5419	6380	6700
Culverin of 15 ,,	650	1430	2860	4290	5150	5720	6180
Demi-culverin of 12 lb.	600	1320	2640	3960	4750	5346	5700
Demi-culverin of 10 ,,	550	1210	2420	3630	4837	5104	5500
Demi-culverin of 8 ,,	500	1100	2200	3300	4125	4640	5000
Saker of 6 lb.	450	990	1980	2970	3742	4176	4500
Falcon of 4 lb.	400	880	1760	2640	3300	3712	4000
Falcon of 2 ,,	330	704	1408	2112	2640	2970	3200

[1] That is, 1st point = $7\frac{1}{2}°$; 2nd point = 15°; 3rd point = $22\frac{1}{2}°$; 4th point = 30°; 5th point = $37\frac{1}{2}°$; 6th point = 45°.

II. *Spanish*: *c.* 1603—(*cont.*)

	Weight of Gun	Weight of Shot	Length of Gun
Demi-cannon	2400 lb.	16 lb.	18 calibres
Culverin	5530	16	30
Demi-culverin	2800	8	26
Saker	1750		
Falconet	1050	3	30

(Diego de Prado y Tovar, *Encyclopaedia de Fundicion de Artilleria y su Platica Manual* (1603), pp. 8 *et seq.*)

III. *English*: *Civil War Period*

	Calibre of Piece[1]	Weight of Piece	Length of Piece	Weight of Shot
Cannon Royal	8 in.	8000 lb.	8 ft	63 lb.
Cannon	7	7000	10	47
Demi-cannon	6	6000	12	27
Culverin	5	4000	11	15
Demi-culverin	4½	3600	10	9
Saker	3½	2500	9½	5¼
Minion	3	1500	8	4
Falcon	2¾	700	6	2¼
Falconet	2	210	4	1¼
Robinet	1¼	120	3	¾

[1] Probably ¼ in. should be allowed for windage. The simplification of the classification and the increased strength of the smaller pieces are worth notice. (William Eldred; *The Gunner's Glasse*, 1646).

IV. *French*: 1666 (de la Fontaine: *Les Fortifications Royales*, 92–).

	Weight of Ball	Calibre	Weight of Gun	Charge	Length of Gun	Point-blank Range	Extreme Range	Gunners and Loaders
Whole Cannon (Flemish)	45 lb.	$49\frac{1}{2}$ lb.	6100 lb.	$22\frac{1}{2}$ lb.	$17\frac{1}{2}$ calibres	1450 ft	16200 ft	5
Whole Cannon (French)	$33\frac{1}{3}$	$36\frac{2}{3}$	5200	17	$19\frac{1}{2}$	1500	16400	5
Demi-cannon (Flemish)	24	$26\frac{1}{2}$	4200	15	$20\frac{1}{2}$	1560	16500	4
Great Culverin (French)	15	$16\frac{1}{2}$	3400	10	33	1630	16625	4
,, ,, ,,	18	20	4000	12	32	1650	16850	4
Quarter Cannon	12	$14\frac{1}{4}$	2800	$8\frac{1}{2}$	$24\frac{1}{2}$	1550	16050	3
Eighth Cannon (Saker)	6	$6\frac{2}{3}$	1700	$4\frac{1}{2}$	29	1500	15925	2
Sixteenth Cannon	3	$3\frac{1}{3}$	1100	$2\frac{1}{2}$	35	1475	15550	2
Thirtysecond Cannon	$1\frac{1}{2}$	$1\frac{13}{20}$	750	$1\frac{1}{2}$	41	1450	15200	2

V. *French*: 1697

	Weight of Piece	Weight of Shot	Length of Piece
French whole cannon	6200 lb.	33 lb.	11 ft
Spanish demi-cannon	5100	24	11 ft
French demi-cannon	4100	16	10 ft. 10 in.
Spanish qtr. cannon	3400	12	10 ft. 9½ in.
French qtr. cannon	1950	8	10 ft. 7½ in.
'Moyenne'	1300	4	10 ft. 7 in.
Falcon	150–800	¼–2	7 ft

The names culverin and saker belong to pieces now disused and 'dont les noms bizarres sont presque inconnus'.

Ordnance of the latest type

24-pr	weighing 3000 lb	and	6 ft. 7·75 in. long
16-pr	2200		6 ft. 2·33 in.
12-pr	2000		6 ft. 1·25 in.
8-pr	1000		4 ft. 11·83 in.
4-pr	600		4 ft. 9 in.

These new pieces had a chamber larger than the bore, were consequently wider at the breech than the old, and cast shorter and lighter.

Ranges at 45°

(Guns of the old type charged with two-thirds of the weight of the ball in powder, the new type with one-third.)

24-pr	4500 yd	8-pr	3320 yd
16-pr	4040	4-pr	3040
12-pr	3740		

'L'on ne s'accorde point sur la Portée des Pièces.'

The windages allowed vary from 0·21 in. for the 33-pr. to 0·11 in. for the 4-pr.

Mortar Ranges

Calibre 12 inch.

Range 240 feet to 2160 feet, elevation 5–45 deg., charge 2 lb.
Increase per deg. 48 feet.

Range 2160 feet to 2700 feet, elevation 36–45 deg., charge 2½ lb.
Increase per deg. 60 feet.

Range 2664 feet to 3240 feet, elevation 37–45 deg., charge 3 lb.
Increase per deg. 72 feet.

Calibre 8 inch:
Range 210 to 1890 feet, elevation 5–45 deg., charge ½ lb.
Increase per deg. 42 feet.
Range 1922 to 2790 feet, elevation 31–45 deg., charge ¾ lb.
Increase per deg. 62 feet.
Range 2870 to 3690 feet, elevation 34–45 deg., charge 1 lb.
Increase per deg. 82 feet.
(Surirey de Saint-Rémy, *Mémoires d'Artillerie* (1697), vol. 1, pp. 57, 60, 70, 77, 258, 261.)

By way of comparison the following table of ranges of naval ordnance in 1820 is given from *A Treatise of Naval Gunnery* by Sir Howard Douglas.

	Point-blank	1°	2°	3°
32-pr., 9 ft. 6 in. long, charge 10–11 lb.	350	750	1050	1320
24-pr., 9 ft. 6 in. long, charge 6 lb.	248	661	847	1213
24-pr., 6 ft. 6 in. long, charge 6 lb.	221	582	832	1133
12-pr., 8 ft. 6 in. long, charge 4 lb.	300	700	913	1189
9-pr., 8 ft. 6 in. long, charge 3 lb.	300	683	900	1200

	4°	5°	6°	7°	8°	9°	10°	
32-pr.	1600	2085	2100	2200	2460	2600	2900	yards
24-pr.	1472	1590	1939	2097[1]	2288	2545	2673	
24-pr.	1308	1545	1741	2273	2250	2204[1]	2562	
12-pr.	1400	1580	1800	—	—	—	—	
9-pr.	1400	1622	1800	—	—	—	—	

[1] misprint?

BIBLIOGRAPHY

A. History of Science and Technology

(1) *Sources*

Abbé Gallon (ed.). *Machines et Inventions approuvées par l'Académie.* Paris, 1735.

Académie Royale des Sciences, Paris. *Histoire de l'Académie . . . depuis* 1666 (to 1699). Continued as *Mémoires.* Paris, 1733.

Divers Ouvrages de mathématique et de physique. Par Messieurs de l'Académie. Paris, 1693.

Acta Eruditorum. Lipsiae, 1682.

Aristotle. *Works* (translation into English, ed. W. D. Ross). Oxford, 1908–31.

Beeckman, Isaac. *Journal* (ed. Cornelis de Waard). 1939– .

Bernoulli, Johann I. *Opera Omnia.* Lausannae et Genevae, 1742.

Birch, Thomas. *History of the Royal Society.* London, 1756–7.

Biringuccio, Vanoccio. *Pirotechnia* (1540). Trans. Smith and Gnudi. New York, 1943.

Boyle, Robert. *Some Considerations touching the Usefulness of Natural Philosophy.* Oxford, 1664.

Works (ed. Thomas Birch) London, 1744, 1772.

Cardano, Girolamo. *Opus novum de proportionibus.* Basileae, 1570.

De Subtilitate Libri xxi. Norimbergae, 1550.

Cavalieri, Bonaventura. *Lo Specchio Ustorio overe Trattato delle Settioni Coniche.* Bologna, 1632.

Descartes, René. *Œuvres* (ed. C. Adam and Paul Tannery). Paris, 1897–1913.

Digges, Thomas. *A Geometrical Treatise named Pantometria.* London, 1571, 1591.

Stratioticos. London, 1590.

Duhamel, J. B. *Regiae Scientiarum Academiae Historia.* Paris, 1698.

Edleston, Joseph. *Correspondence of Sir Isaac Newton and Professor Cotes.* London, 1850.

EULER, LEONHARD. *Neue Grundsätze der Artillerie aus dem Englishchen des Herrn Benjamin Robins ubersetzt und mit vielen Anmerkungen versehen.* Berlin, 1745.

Opera Omnia. Leipzig and Berlin, 1922. (English trans. 1777, French trans. 1783.)

FABRY, HONORÉ. *Tractatus Physicus de Motu Locali.* Lyons, 1646.

FOURNIER, GEORGES. *Hydrographie.* Paris, 1643, 1667.

GALILEI, GALILEO. *Discorsi e Dimostrazioni matematiche intorno a due nuove scienze.* Leide, 1638.

Opere. Edizione Nazionale, Firenze, 1890.

GERHARDT, K. I. *Leibnizens mathematische Schriften* (in *Gesammelte Werke*, hrsg. von G. H. Pertz). Hanover, Berlin, Halle, 1843-63.

GREGORY, JAMES. *Tentamina quaedam Geometrica de Motu Penduli & Projectorum.* (See Mathers, Patrick.)

GREW, NEHEMIAH. *Musaeum Regalis Societas.* London, 1681.

GUNTHER, ROBERT. *Early Science in Oxford.* Oxford, 1902-45.

HALE, THOMAS. *An Account of Several New Inventions and Improvements.* London, 1691.

HALLEY, EDMUND. 'A discourse concerning Gravity, and its properties, wherein the descent of heavy bodies and the motion of projects is briefly but fully handled; together with the solution of a problem of great use in Gunnery.' *Phil. Trans.* vol. XVI (1686-7).

'A Proposition of general use in the art of Gunnery, showing the rule of laying a mortar to pass, in order to strike any object above or below the horizon.' *Ibid.* vol. XIX (1695-7).

HERMANN, JACOB. *Phoronomia, sive de viribus et motibus corporum solidorum et fluidorum libri duo.* Amstelaedami, 1716.

HOOKE, ROBERT. *Diary, 1672-1680.* London, 1935.

HUYGENS, CHRISTIAAN. *Œuvres Complètes.* La Haye, 1885-1950.

KÜNCKEL VON LOWENSTERN, JOHANN. *Collegium Physico-chemicum Experimentale oder Laboratorium Chymicum.* Hamburg and Leipzig, 1716.

LEIBNIZ, G. G. *Œuvres . . . publiées pour la première fois* (ed. L. A. Foucher de Careil), Paris, 1859-65.

MACPIKE, E. F. *Correspondence and Papers of Edmond Halley.* Oxford, 1932.

MARIOTTE, EDMÉ. *Œuvres*, Leyden, 1717.

MATHERS, PATRICK (i.e. William Saunders). *The Great and New Art of Weighing Vanity*. Glasgow, 1672. (Cf. p. 121 n. 2.)

MERSENNE, MARIN. *Les Questions Théologiques, Physiques, Morales et Mathématiques*. Paris, 1634.

Harmonie Universelle. Paris, 1636-7.

Universae Geometriae Mixtaeque Mathematicae Synopsis. Paris, 1644.

Cogitata Physico-Mathematica. Paris, 1644.

NEWTON, SIR ISAAC. *Philosophiae Naturalis Principia Mathematica*. London, 1687, 1713, 1726.

PETTY, SIR WILLIAM. *The Petty Papers* (ed. Marquis of Lansdowne). London, 1927.

The double-bottom or twin hulled ship of Sir William Petty. Oxford, 1931.

Philosophical Transactions of the Royal Society. London, 1665- .

PLATTE, SIR HUGH. *The Jewell House of Art and Nature*. London, 1594.

PLATTES, GABRIEL. *A Discovery of Infinite Treasure hidden since the World's Beginning*. London, 1639.

PLOT, ROBERT. *Natural History of Staffordshire*. London, 1686.

PORTA, GIAMBATTISTA DELLA. *Natural Magick*. London, 1658.

RÉAUMUR, R. A. FERCHAULT DE. *L'Art d'Adoucir le Fer Fondu*. Paris, 1722.

RICCIOLI, GIOVANNI BATTISTA. *Almagestum Novum, Astronomian veterem novamque complectens*. Bologna, 1651.

RIGAUD, S. J. *Correspondence of Scientific Men of the seventeenth century ... printed from the originals in the collection of the Earl of Macclesfield*. Oxford, 1841.

RŒMER, OLE. *Adversaria* (ed. Thyra Eibe and Kirstine Meyer). Copenhagen, 1910.

Royal Society MSS, Journal Books, Register Books, Classified Papers.

SANTBECH, DANIEL. *Problematum astronomicorum et geometricorum sectiones septem*. Basileae, 1561.

SPRAT, THOMAS. *History of the Royal Society*. London, 1667.

STURMY, SAMUEL. *Mariner's Magazine*. London, 1669, 1684.

SWEDENBORG, EMMANUEL. *Regnum Subterraneum sive minerale de Ferro*. Dresden und Leipzig, 1734.

TACQUET, ANDRÉ. *Opera Mathematica*. Antwerp, 1669.

Tartaglia, Niccolo. *La Nova Scientia.* Venice, 1537.
Quesiti et Inventioni Diverse. Venice, 1546, 1551.

Torricelli, Evangelista. *Opere.* Faenza, 1919.
Opera Geometrica, Florentie, 1644.

Turnbull, H. W. *James Gregory Tercentenary Volume.* London, 1935.

Waard, Cornelis de (ed.) *Correspondance du P. M. Mersenne, religieux Minime.* Paris, 1932.

Waller, Richard. *Essayes of Natural Experiments made in the Accademia del Cimento.* London, 1684.

Wallis, John. *Opera Mathematica.* Oxford, 1695.

Webster, John. *Metallographia.* London, 1671.

Wilkins, John. *Mathematical Magick.* London, 1648.

(2) *Secondary Works*

Artz, F. B. 'Les Débuts de l'Éducation Technique en France 1500-1700', *Revue d'Histoire Moderne*, vol. xii, 1937.

Beck, Ludwig. *Die Geschichte des Eisens in Technischer und Kulturgeschichtlicher Berziehung.* Braunschweig, 1884-1903.

Blackman, Herbert. 'Gunfounding at Heathfield in the 18th century.' *Sussex Archeological Collections*, vol. lxvii. Cambridge, 1926.

Charbonnier, P. *Essais sur l'Histoire de la Balistique.* Paris, 1928.

Clark, G. N. *Science and Social Welfare in the Age of Newton.* Oxford, 1937.

Dircks, H. *The Life, Times and Scientific Labours of the second Marquis of Worcester.* London, 1865.

Duhem, Pierre. *Études sur Léonard de Vinci.* Paris, 1906-13.

Isis. Ed. George Sarton. 1913- .

Jenkins, Rhys. *Collected Papers.* Cambridge, 1936.
'The Rise and Fall of the Sussex Iron Industry.' *Transactions of the Newcomen Society*, vol. i, 1920.

Johnson, F. R. *Astronomical Thought in Renaissance England.* Baltimore, 1937.

Jones, R. F. *Ancients and Moderns.* St Louis, 1936.

Koyré, Alexandre. *Études Galiléennes. Actualités Scientifiques et Industrielles* 852. Paris, 1939.

Lenoble, Robert. *Mersenne ou la Naissance du Mécanisme.* Paris, 1943.

Mach, Ernst. *The Science of Mechanics* (English trans., 3rd ed.). Chicago, 1907.

Maury, F. *Les Académies d'autrefois.* Vol. I, *L'ancienne académie des Sciences.* Paris, 1864.

Medieval Studies. New York, 1941- .

Merton, Robert K. 'Science, Technology and Society in Seventeenth Century England.' *Osiris*, vol. IV, 1938.

Montucla, J. E. *Histoire des Mathématiques.* Paris, 1758, 1799-1802.

More, Louis Trenchard. *Isaac Newton.* New York, 1934.

Napier, Mark. *Memorials of John Napier of Merchistoun.* Edinburgh, 1834.

Parsons, J. L. 'The Sussex Ironworks'. *Sussex Archeological Collections*, vol. XXXII. Lewes, 1882.

Sarton, George. *Introduction to the History of Science.* Baltimore, 1927- .

Schubert, H. 'The first Cast Iron Cannon made in England'. *Journal of the Iron and Steel Institute*, vol. CXLVI. London, 1943.

Science at the Crossroads. London, 1931.

Straker, Ernest. *Wealden Iron.* London, 1931.

Tallqvist, H. J. *Översikt av Ballistikens Historia* (Svenska Tekniska Vetenskapsakademien i Finland, Acta, Band IX.) Helsingfors, 1931.

Tannery, Paul. *Mémoires Scientifiques.* Paris, 1912- .

Targione-Tozzetti, G. *Notizie degli Aggrandimenti Della Scienze Fisiche accaduti in Toscani.* Firenze, 1780.

Thorndike, Lynn. *History of Magic and Experimental Science.* London, 1923- .

Thorpe, W. H. 'The Vauxhall Ordnance Factory of Charles I.' *Transactions of the Newcomen Society*, vol. XIII, 1932-3.

Wohlwill, Emil. 'Die Entdeckung des Beharrungsgesetzes.' *Zeitschrift für Volkerpsychologie*, vols. XIV, XV.

Wolf, Abraham. *History of Science, Technology and Philosophy in the 16th and 17th centuries.* London, 1935.

Woodcroft, Bennet. *Titles of Patents of Invention chronologically arranged.* London, 1854.

Zenghelis, C. 'Le Feu Grégois et les Armes à Feu des Byzantins.' *Byzantion* vol. VIII, 1932.

B. Military and Naval History

(1) *Sources*

ALABA Y VIAMONT, DIEGO DE. *El Perfeto Capitan instruido en la disciplina Militar y nueva ciencia de la Artilleria.* Madrid, 1590.

ALBEMARLE, GEORGE MONK, DUKE OF. *Observations upon Military and Political Affairs.* London, 1671.

ALBERGHETTI, SIGISMONDO. *Nova Artilleria Veneta.* Venice, 1703.

ANDERSON, ROBERT. *The Genuine Use and Effects of the Gunne.* London, 1674.

BABINGTON, JOHN. *Pyrotechnia, or a Discourse of Artificial Fireworks.* London, 1635.

BELIDOR, B. FOREST DE. *Le Bombardier François.* Paris, 1731.

BINNINGS, THOMAS. *A Light to the Art of Gunnery.* London, 1675.

BLONDEL, F. *L'Art de Jetter les Bombes.* Paris, 1683.

BOURNE, WILLIAM. *Inventions or Devises very necessary for all Generalles and Captaines or Leaders of men, as well by Sea as by Land.* London, 1578.

The Arte of Shooting in Great Ordnaunce. London, 1587.

BUSCA, GABRIEL. *Della Espugnatione et Difesa della Fortezze.* Turino, 1598.

COLLADO, LUYS. *Prattica Manuale dell'Artiglieria.* Milano, 1606.

CORBETT, JULIAN S. *Fighting Instructions 1530–1816.* N.R.S. 1905.

DAVIES, EDWARD. *The Art of War and England's Traynings.* London, 1619.

Egerton Papers. Camden Society, 1840.

ELDRED, WILLIAM. *The Gunner's Glasse.* London, 1646.

LA FONTAINE, —— DE. *Les Fortifications Royales.* Paris, 1666.

FURTENBACH, JOSEPH. *Halinitro-pyrobolia. Beschreibung einer neuen Buchsenmeisterey.* Ulm, 1627.

GAYA, LOUIS DE. *Traité des Armes, des machines de Guerre, des feux d'artifices.* Paris, 1678.

GOLDMAN, NICHOLAS. *Elementorum Architectura Militaris Libri iv.* Leyden, 1643.

LUCAR, CYPRIAN. *Three Bookes of Colloquies.* London, 1588.

MARKHAM, FRANCIS. *Five Decades of Epistles of Warre.* London, 1622.

MEYERSTEIN, E. H. W. *Adventures by Sea of Edward Coxere.* Oxford, 1945.

MONSON, SIR WILLIAM. *Naval Tracts.* N.R.S. 1913.

MOORE, SIR JONAS. *Modern Fortification, or Elements of Military Architecture.* London, 1689.

(trans.) *A General Treatise of Artillery and Great Ordnance. Writ in Italian by Tomaso Moretti of Brescia.* London, 1683.

NICHOLAS, SIR HENRY. *Despatches and Letters of Lord Nelson.* London, 1845.

NORTON, ROBERT. *The Gunner, shewing the whole Practise of Artillerie.* London, 1628.

NYE, NATHANIEL. *Art of Gunnery.* London, 1647.

ORRERY, ROGER, EARL OF. *Treatise on the Art of War.* London, 1677.

PEPYS, SAMUEL. *Naval Minutes.* N.R.S. 1925.

Tangier Papers. N.R.S. 1935.

Perrin, W. G. (ed.). *Boteler's Dialogues.* N.R.S. 1929.

PRADO Y TOVAR, DIEGO DE. *Encyclopaedia de Fundicion de Artilleria y su Platica Manual* (MS.). 1603.

Public Record Office, War Office Records, especially *Ordnance Minutes,* W.O. 47.

RIVAULT DE FLEURANCE, DAVID. *Les Elemens de l'Artillerie.* Paris, 1605.

ROBINS, BENJAMIN. *New Principles of Gunnery.* London, 1742.

Mathematical Tracts. London, 1761.

ROBERTS, JOHN. *The Complete Cannoniere.* London, 1639.

RYFF, WALTER. *Der Furnembsten notwendigsten der gantzen Architectur.* Nürnberg, 1547.

SAINT-RÉMY, SURIREY DE. *Mémoires d'Artillerie.* Paris, 1697.

SIEMENOWICZ, CASIMIR. *Artis Magnae artilleriae pars prima.* Amsterdam, 1650.

SMITH, JOHN. *An Accidence, or the pathway to experience necessary for all Young Seamen.* London, 1626.

Also as: *Seaman's Grammar.* London, 1653.

Seaman's Grammar and Dictionary. London, 1691.

UFFANO, DIEGO. *Artillerie: c'est à dire Vraye Instruction de l'Artillerie et de toutes ses appurtenances* (trans. Th. de Brye). Frankfort, 1614.

VENN, THOMAS. *Military and Maritine Discipline.* London, 1672.

WARD, NED. *The Wooden World Dissected.* London, 1706.

WARD, ROBERT. *Anim' Adversions of Warre.* London, 1639.

WHITEHORNE, PETER. *Certain Waies for the Orderyng Souldiers in Battelray and setting of Battailes.* London, 1562.

ZUBLER, LEONHARD. *Nova Geometria Pyrobolia. Neuwe Geometrische Buchsenmeisteren.* Zurich, 1614.

(2) *Secondary Works*

Cruikshank, C. G. *Elizabeth's Army.* London, 1946.

Demmin, Auguste. *Weapons of War* (English trans.). London, 1870.

Douglas, Sir Howard. *Naval Gunnery.* London, 1817, 1820, 1860.

Duncan, F. *History of the Royal Regiment of Artillery.* London, 1872.

ffoulkes, Charles. *The Gunfounders of England.* Cambridge, 1937.

Firth, Sir Charles. *Cromwell's Army.* London, 1902.

George, J. N. *English Guns and Rifles.* Plantersville, 1947.

Goodrich, L. Carrington and Feng Chia Sheng. 'Early Chinese Fire-arms.' *Isis*, vol. XXXVI, 1946.

Greener, William. *The Gun.* London, 1835.

Grose, Francis. *Military Antiquities.* London, 1786.

Hime, H. W. L. *Gunpowder and Ammunition: their Origin and Progress.* London, 1904.

Jackson, H. J. *European hand firearms of the 16th, 17th, and 18th centuries.* London, 1923.

Jähns, M. *Geschichte der Kriegswissenshaften vornehmlich in Deutschland.* München, 1889.

Journal of the Society for Army Historical Research. London, 1921.

La Roncière, Charles de. *Histoire de la Marine Française.* Paris, 1899- .

Lazard, P. *Vauban, 1633-1707.* Paris, 1934.

Lewis, Michael. 'Armada Guns.' *Mariner's Mirror*, vols. XXVIII, XXIX, 1942-3.

Lloyd, E. W. and Hadcock, A. G. *Artillery: its Progress and Present Position.* Portsmouth, 1893.

Martin-Leake, Stephen. *Life of Captain Stephen Martin* (ed. C. R. Markham). N.R.S. 1895.

Life of Sir John Leake (ed. Geoffrey Callender). N.R.S. 1918.

Napoleon III and Col. Favé. *Études sur le passé et l'avenir de l'artillerie.* Paris, 1846-63.

Nef, J. U. 'War and Economic Progress 1540–1640.' *Economic History Review*, vol. XII, 1942.

Oman, Sir Charles. *The Art of War in the Middle Ages*. London, 1924.
The Art of War in the Sixteenth century. London, 1937

Oppenheim, M. *A History of the Administration of the Royal Navy 1509–1600*. London, 1896.

Pollard, H. B. C. *History of Firearms*. London, 1926.

Poten, B. 'Geschichte des Militär Erziehungs und Bildungswesens in den Landen deutscher Zunge.' *Monumenta Germaniae Paedagogica*, vols. X *et seq*. Berlin, 1889.

Proceedings of the Royal Artillery Institution (cont. as *Journal of the Royal Artillery*).

Raikes, G. A. *History of the Honourable Artillery Company*. London, 1878.

Robertson, F. L. *Evolution of Naval Armament*. London, 1921.

Tout, T. F. 'Firearms in England in the 14th century.' *English Historical Review*, vol. XXV, 1911.

Walton, Clifford. *History of the British Standing Army 1660–1700*. London, 1894.

Young, H. A. *The East India Company's Arsenals*. Oxford, 1937.

C. General

Acts of the Privy Council. London, 1890– .

Boissonade, P. *Colbert*, 1661–1683. Paris, 1932.

Calendar of Patent Rolls. London, 1901– .

Calendar of State Papers, Domestic. London, 1856– .

Cellini, Benvenuto. *Vita* (various ed. and trans.).

Evelyn, John. *Diary* (ed. H. B. Wheatley). London, 1906.

Hatfield, W. H. *Cast Iron*. London, 1918.

Historical Manuscripts Commission; De l'Isle and Dudley; Earl Cowper; 13th Report, Appendix, part IV.

Notestein, Wallace *et al*. *Commons' Debates of* 1621. Newhaven 1935.

Pepys, Samuel. *Diary* (ed. H. B. Wheatley). London, 1946

Ralegh, Sir Walter. *Observations touching Trade and Commerce*. London, 1653.

RAY, JOHN. *Observations Topographical.* London, 1673.
A Collection of English Words not generally used, etc. London, 1691.

TAWNEY, R. H. and POWER, E. E. *Tudor Economic Documents.* London, 1924.

WARBURTON, ELIOT. *Memoirs of Prince Rupert and the Cavaliers.* London, 1849.

INDEX